Dispersed Manufacturing Networks

Rob Dekkers
Editor

Dispersed Manufacturing Networks

Challenges for Research and Practice

Springer

Editor
Rob Dekkers, PhD
Div. Management & Business Economics
Paisley Business School
PA1 2BE Paisley
UK
dek.rob@gmail.com

ISBN 978-1-4471-5755-7 ISBN 978-1-84882-468-3 (eBook)
DOI 10.1007/978-1-978-1-84882-468-3
Springer Dordrecht Heidelberg London New York

British Library Cataloguing in Publication Data
A catalogue record for this book is available from the British Library

Cover design: eStudioCalamar, Figueres/Berlin

Printed on acid-free paper

Springer is part of Springer Science+Business Media (www.springer.com)

TO NIL AND MERT

Acknowledgements

Many have contributed to this edited book, most of all the authors of the respective chapters. Without their research and their efforts and the intense correspondence that lead to the stage where we are now, this book on *Dispersed Manufacturing Networks*, would not have been possible. The authors always showed understanding by their quick and adequate responses to numerous requests. It was a privilege to work with the authors and exchange thoughts on this topic that is at the forefront of academic research.

Secondly, I would like to acknowledge my gratitude to David Bennett, editor of the *Journal of Manufacturing Technology Management*, and Emerald Group Publishing, the publisher of that journal, for their permission to build on the papers of the *Special Issue on Dispersed Manufacturing Networks*, which appeared in 2006 (Volume 17, Issue 8); this concerns Chapters 3, 6, 7, 8, 10 and 11. All these earlier journal papers have been amended to account for the progress made during the two years after the Special Issue. Two chapters that did not make it to Special Issue, mostly due to the pressing deadlines, complement the rewritten versions of the journal papers; these chapters (4 and 9, based on the earlier submissions) help to complete the picture of *Dispersed Manufacturing Networks* and provide additional insight. Furthermore, it should be noted that an earlier version of Chapter 2 appeared in the proceedings of the *12th International EurOMA Conference*, held in Budapest (2005); the discussions at that conference inspired the writing of this chapter as it is now. Moreover, an earlier version of Chapter 5 appeared in the proceedings of the *9th Annual Cambridge International Manufacturing Symposium*, Cambridge (2004), again encouraged by remarks of the participants.

Thirdly, my indebtedness extends to the reviewers, particularly for the efforts they made for appraising all possible contributions to the Special Issue of the *Journal of Manufacturing Technology Management*, which in turn formed the base for the writings in the book. All chapters, except Chapter 2 and 5, were initially reviewed by: Luis Camarinha-Matos, Vittorio Chiesa, Martin Christopher, David Bennett, Angappa Gunasekaran, Ian Hipkin, Bernard Hon, Thanos Kourouklis, Hermann Kühnle, Sameer Kumar, W.B. Lee, Bart MacCarthy, Mohamed Naim, Andrew Nee, Günther Schuh, Jongyiang Shi, Ann Vereecke, Jack van der Vorst, Seungjin Wang, and Zhang Shu. The critiques were constructive and extensive allowing the authors to reshape the chapters and to take in consideration the valuable comments in addition to my reviewing and remarks.

Fourthly, I credit Carole Gould, Shishank Shishank, Kumaran Sugumaran and Julie Thomson, who have all assisted with the final proofing of the manuscripts. Particularly, Kumaran Sugumaran and Shishank Shishank have taken the time to do the proofing for the majority of the chapters.

Finally, I would like to thank Springer UK (Anthony Doyle, Claire Protherough and Simon Reese) and their editor of the *Advanced Manufacturing Series*, Duc - Truong Pham, for the opportunity to compile and publish this edited book.

Should I have forgotten anyone who has contributed or provided valuable assistance, they should consider this an acknowledgement of their effort and an apology at the same time.

Rob Dekkers *Paisley, 31 December 2008*

Contents

Acknowledgements v

PROLOGUE 1

1 Introduction 3

2 Industrial Networks of the Future: Review of Research and Practice 13
Rob Dekkers and David Bennett

PART I: Networks as Complex Adaptive Systems 35

3 Dispersed Network Manufacturing: An Emerging Form of Collaboration Networks 39
Hamid Noori and W.B. Lee

4 Self-Similarity and Criticality in Dispersed Manufacturing: A Contribution to Production Networks Control 59
Hermann Kühnle

5 Collaborations in Industrial Networks: The Co-Evolutionary Perspective 77
Rob Dekkers

PART II: Control and Coordination in Industrial Networks 107

6 Designing and Modeling Agile Supply-Demand Networks 111
Petri Helo, Natalia Kitaygorodskaya, Sari Salminen and Roger J. Jiao

7 Framework for Developing an Agile Future-Proof Supply Chain 131
Hossein Sharifi, Hossam Ismail and Iain Reid

PART III: International Issues of Industrial Networks 155

8 Developing a Worldwide Production Network 159
Joachim Kuhn

9 Planning in Companies with Dispersed Capacity 179
Stephen A. Smith, David J. Petty, David G. Trustrum and Ashraf W. Labib

10 Managing Product Variety in Multinational Corporation Supply Chains: A Simulation Study Investigating Flow Time 195
Mahendrawathi Er and Bart MacCarthy

11 Set-Up and Operation of Global Engineering Networks Spanning Industrialized and Emerging Economies 223
Harshavardhan Karandikar

EPILOGUE 241

12 What Follows 243

About the Editor 251

About the Authors 253

PROLOGUE

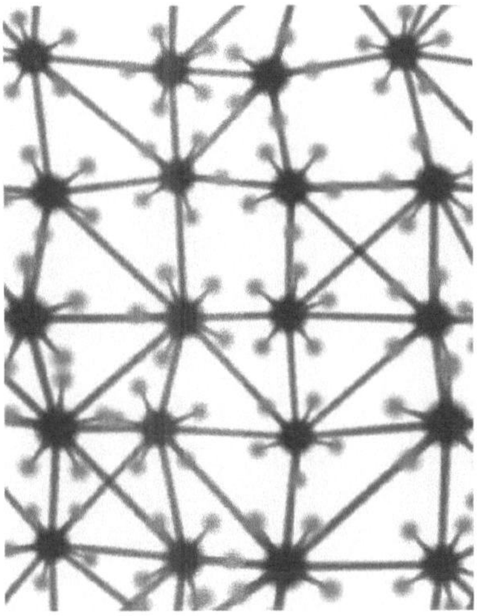

1 Introduction

Historically, networks have existed for a long time. It suffices to point to the *Silk Route* as an ancient example of the global supply chain or to the existence of trading between Asia and Europe by the Dutch *Vereenigde Oostindische Compagnie* during the Golden Age of the Republic of the Netherlands (16th and 17th centuries), see Dekkers (2005, p. 14). Even then, the contextual environments, i.e. the social environment in which the networks existed, determined for a large part the transactional environment of trading relationships. Social-economic historians have investigated this domain to understand the networks that were present during the Commercial Revolution in the Middle Ages, an era that saw the resurgence of Mediterranean and European long-distance trading (e.g. Greif, 1996). Later, the global supply chains, focusing on basic needs, agricultural goods and raw materials, were affected by the First Industrial Revolution (Brasseul, 1998, p. 8). Firstly, the growing demand during that period increased the volume of trade. Secondly, the capability of sources (regions and nations) to produce their own intermediaries or products preluded the emergence of industrial networks. This only tells us that during the Second and Third Industrial Revolutions, trade and industry increasingly relied on the networks they created to sustain competitive advantage.

Taking advantage of the competitive offering of products (and services) to customers in geographically dispersed, emerging, and established global markets nowadays demand higher quality products of a greater variety and at a lower cost with shorter response times. As a result, firms have been forced to reorganise their activities and realign their global strategies in order to provide the speed and flexibility necessary to respond to windows of market opportunity (Dekkers and van Luttervelt, 2006, pp. 4-6). Consequently, organisations have moved from centralised, vertically integrated, single-site manufacturing facilities to geographically dispersed networks of resources, a trend already noted by Miles and Snow (1984). Additionally, in order to acquire technological know-how and assets quickly, or to acquire a local presence in new and distant markets, strategic partners are increasingly part of the network structure (Hemphill and Vonortas, 2003, p. 255). These changes require adaptations by companies to fit the characteristics of industrial networks in often-dynamic environments.

1.1 Traditional Focus of Research

Within the domain of industrial networks, many studies have preceded this book in outlining prospects for resolving these challenges for industry and for issues in research (e.g. Camarinha-Matos and Afsarmanesh, 2005; Gulati *et al.*, 2000; Karlsson, 2003). Camarinha-Matos and Afsarmanesh add that a discipline of Collaborative Networks should focus on the structure, behaviour and evolving dynamics of autonomous entities that collaborate to better achieve common or compatible goals. This denotes two essential issues for networks: autonomous entities and collaboration; these issues have received less attention in the context of more loosely connected entities.

There are already many perspectives from which to look at the structure and dynamics of collaborations between autonomous agents in networks. This section presents three of these perspectives: (i) the Strategic Network Model – as representative of approaches residing in the monolithic firm –, (ii) the Resource-Based View – as a popular foundation for explaining inter-firm relationships - and (iii) the social dynamics of inter-organisational relationships – as an extensively researched issue –, since they have dominated strategic thinking about industrial networks. Two of these three perspectives presume that networks are driven by strategic objectives of a (often) dominant partner, as will be demonstrated; the third perspective - the social dynamics of inter-organisational relationships – preludes the discussion of Dispersed Manufacturing Networks in Section 1.2 as networks of more loosely connected entities that seek collaboration and hence, differ from the two dominant perspectives: the Strategic Network Model and the Resource-Based View.

Strategic Network Model

The first perspective defines Strategic Networks as long-term, purposeful arrangements among distinct, but related, for-profit organisations that allow members to gain or sustain competitive advantage over their competitors outside the arrangement (Jarillo, 1988, p. 32). Kogut (2000) describes this type of network for the Toyota Production System. In this view, Strategic Networks are merely a superior method of managing the process necessary for the production and sale of a chosen set of products (like in Freiling [1998]). It should be noted that some associate the term Strategic Networks with the concept of networked organisations in general (e.g. Gulati *et al.*, 2000) and some with supply chains (e.g. New and Mitropoulos, 1995). Levin (1998) discusses the advantages of networks for small businesses, which help them to create a greater reach. But most of all, (strategic) alliances and joint-ventures typically exemplify these Strategic Networks. According to Gulati (1998, pp. 294–295), studies have focused on (a) the inclination of firms to enter alliances and the formal contracts, (b) the formation and performance of alliances and (c) the firm- and industry-level factors that impel organisations to enter alliances. Still less attention has been paid to the management of these arrangements. However, it appears that the factors power and trust dominate these types of network relationships (Das and Teng, 2001; Huemer, 2004; Thorelli, 1986, p. 38). Hence, these Strategic Networks came into existence through strategic objectives of one or more of the partners, which make it necessary to collaborate and which create tensions in inter-organisational relationships.

Such Strategic Networks, as purposeful arrangements (e.g. Gulati *et al.*, 2005), hardly address the issues of autonomous agents. Even alliances, which everybody generally perceives as more stable relationships between firms, dissolve over time or end up in mergers, according to a study into 92 alliances by Kogut (1989). Consequently, if the balance shifts to independence of agents, which is dependent on the uniqueness of their resources, the network will perform local optimisation and will create power shifts (Medcof, 2001). Henceforth, the need for more loosely connected agents and the flexibility to capture market opportunities undermines the arrangements of Strategic Networks, resulting in issues of power and trust in the relationships (Vangen and Huxham, 2000). That way, instability seems to dominate the continuity and constituency of industrial networks, depending on the uniqueness of their resources; ultimately, the Strategic Network, as purposeful arrangement, is driven by the strategic objectives of the dominant partner.

Resource-Based View

The Resource-Based View, as similar thinking, quickly followed the concept of Strategic Networks in the 1980s; although this view can be traced back to the 1960s and 1970s when organisational theorists combined research on inter-organisational relations and the political economy of organisations (Hemphill and Vonortas, 2003, p. 261). Later, resources were defined as tangible and intangible assets that are tied semi-permanently to a firm (Wernerfelt, 1984, p. 172). Others, especially Barney (1991), have articulated this view by shifting the emphasis from organisational theory to the organisation's goal of reducing the uncertainty and the dependency on other organisations for its survival. To confer competitive advantage, resources must not be possessed by all competing firms, they must be difficult to imitate or duplicate through other means (Barney *et al.*, 2001, p. 625) and contribute positively to performance; Hamel and Prahalad (1994) follow similar reasoning for the concept of core competencies. Contrarily, the resource dependencies among organisations constrain their behaviour and ultimately translate into power differentials that must be effectively controlled. Hoopes *et al.* (2003, p. 897) remark that the Resource-Based View originates in simple applications of micro-economics, industrial organisation, organisational theory and traditional business policy and that it would be better treated in the context of competitive heterogeneity (a view shared by Dyer and Singh [1998]). By using this perspective, a network can develop and exploit a set of resources, such as knowledge, technology and organisational skills, for specific products (Grant, 1996, pp. 119–120) and market positioning to capture market opportunities (Chakravarty, 2001). The network strategy must not only focus on the exploitation of the competitive advantage but also on the utilisation and acquisition of resources to create that advantage; therefore, the network strategy extends beyond the strategic intents of the (dominant) monolithic company.

Social Dynamics of Inter-Organisational Relationships

But this raises questions to examine the third perspective on industrial networks: the social dynamics of relationships. Within the mindset of this interpretation, Uzzi (1997, pp. 36–37) refers to other research that has shown that network relationships in the Japanese automotive industry (for example, Toyota [Dyer and Nobeoka, 2000]) and the Italian knitwear industries are characterised by trust and personal ties, rather than by explicit contracts. Additionally, he points to investigations that reveal that embedded actors in regional production networks satisfice rather than maximise on price; they shift their focus from the narrow economic goal of winning gain and exploiting dependence to cultivating long-term, cooperative ties (similar findings appear in Hardy *et al.* [2003], who investigated a Palestinian non-governmental organisation). Kaufman *et al.* (2000, p. 660) arrived at a similar conclusion: collaborative relationships pay off for the supplier. Their findings suggest that suppliers should focus either on low-cost strategy or on a collaborative partnership to yield most profits. Uzzi (1997, p. 54, 61) found that embeddedness assists adaptation because actors can better identify and execute coordinated solutions to organisational problems. In networks of close ties, motivation is neither purely selfish nor cooperative (in contrast to arm's-length relations) but an emergent property of the social structure within which actors are embedded.

1.2 Dispersed Manufacturing Networks

This observation reinforces the view that contemporary industrial networked structures are not only rooted in the monolithic company; they also come about through the intense cooperation and collaboration between, often smaller, firms as actors in (the social structure of) networks. Bennett and Dekkers (2005) have identified this as one of the main drivers for the rise of industrial networks. The so-called "Third Italy" can be seen as an exponent of this type of collaboration between these firms at a regional level (e.g. Biggiero, 1999; Robertson and Langlois, 1995, p. 549), see Figure 1.1. On these dimensions, Strategic Networks and networks evolving from the Resource-Based View score high on ownership integration (e.g. holding companies and the Chandlerian firm). However, contemporary industrial networks rely less on ownership but require some degree of collaboration and coordination.

This brings us to the topic of this book: *Dispersed Manufacturing Networks* as an organisational manifestation for collaboration between and coordination across loosely connected agents. Industrial networks provide an answer to the current challenges of innovative potential, responsiveness and flexibility through their capability to absorb change and to capture market opportunities. The emerging possibilities of information technology and data-communication, the globalisation of markets, and the ongoing specialisation of firms have paved the way. All these simultaneous developments foster the specific characteristics of (international) networks of companies: collaboration for manufacturing and delivery of products and services, decentralisation of decision-making, and inter-organisational integration for co-ordination (adapted from O'Neill and Sackett [1994, p. 42]); Dispersed Manufacturing Networks harbour these in themselves often conflicting characteristics.

In this perspective, Dispersed Manufacturing Networks – Lee and Lau (1999) have attributed this label to agile manufacturing networks of loosely connected entities; they comprise the total primary process, product development and supply chain, in an international setting, relying on applications of information technology and data-communication to exchange information and to coordinate actions. The quest for agility in Dispersed Manufacturing Networks indicates the capability to operate virtually at the borderline between outsourcing (as make-or-buy) and alliances (as a continuous form of cooperation between partners). This has profound

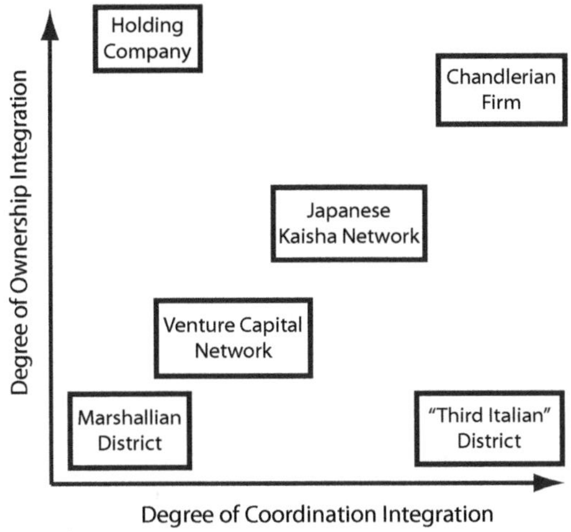

Figure 1.1. *Two dimensions of integration, according to Robertson and Langlois (1995, p. 548). The concept of the holding company (as conglomerate of "independent" business units) and the Chandlerian firm indicate the traditional monolithic company. The Japanese Kaisha and Keiretsu networks caused an upheaval during the 1980s for challenging arms-length practices in Western industry. Current developments for industrial networks point to a lower degree of ownership whilst calling on various degrees of coordination (Venture Capital Networks, Marshallian District, "Third Italian" District).*

implications for the resource allocation processes. Firstly, every opportunity requires tuning of available resources to match the specific demands. Secondly, optimisation of resource allocation becomes an issue that exceeds a singular opportunity. Thirdly, agents act independently and therefore the tuning of resources becomes a two-way communication issue for the network. This means without doubt that decision-making for resource allocation becomes anything but a one-time decision or that allocation decisions become fixed through establishing alliances. The continuous motion within industrial networks makes those decisions hold only for temporary states, accounting for contingencies and the assessment of each opportunity.

However, the bulk of available works is devoted to the contractual aspects and the social dynamics of inter-organisational relationships (Nassimbeni, 1998, p. 539); the dynamic forms of communication and coordination have been neglected (accounting for the three characteristics: collaboration, decentralisation and inter-organisational integration). And, therefore, they require more attention from researchers in the domain of industrial networks. Additionally, to deal with the specific characteristics of industrial networks research can yield theories and approaches in two ways: (a) by converting existing theories for individual firms to the domain of networks, and (b) by developing new theories from new perspectives. Converting existing theories makes sense for guided networks (the domain of Strategic Networks, e.g. Supply Chain Management). The true advantages of networked firms, the decentralisation within the network and the specialisation of individual agents, will only be partially addressed in guided networks. The more deterministic methodologies that might fit the domain of individual organisations in guided networks need to be expanded to the non-linear characteristics of networks, in which interaction amongst agents dominates. Yet, there is a strong need for adequate, integrative approaches to manage the collaboration, decentralisation, and the inter-organisational integration as a consequence of these loosely-connected entities, often operating in an international context. Therefore, the aim of this book is to yield methodologies, methods, and tools for industrial networks that are more appropriate (query raised in Dekkers and van Luttervelt [2006, p. 4]), based on the main questions:

- Which forms of industrial networks have emerged and do they meet the industrial challenges of this century? Which performance criteria do they meet?
- Which models will lead to methods and tools that apply for networks? What about the coordination and inter-organisational integration in networks of loosely connected agents?
- How should companies manage industrial networks given their instability and the dynamics of the environment? Which routes does management science have to explore?

1.3 Outline of Book

This book builds on separate research (projects) that address the challenges for collaboration, decentralisation and inter-organisational integration as witnessed in Dispersed Manufacturing Networks – seen as a manifestation of loosely connected entities in a Collaborative Network. The contributions to this book have been separated into four groups: an introductory chapter and three parts.

Industrial Networks of the Future

The introductory contribution to this book by Rob Dekkers and David Bennett (Chapter 2) looks at the industrial networks of the future. It presents an approach for industrial networks of the future, i.e. for the next 15 years and beyond, based on ongoing research and additional considerations. It takes as starting point that the stage has not yet been reached where networks are configured optimally and network operations have reached a stage of maturity. Consequently, this chapter also aims at outlining a research agenda.

Part I: Networks as Complex Adaptive Systems

The following three contributions in Part I approach industrial networks as complex systems. The contribution by Hamid Noori and W.B. Lee in Chapter 3 offers two perspectives. The first one is approaching Dispersed Manufacturing Networks as an archetype for how smaller firms may compete on a global scale. The second one viewis Dispersed Manufacturing Systems as Complex Adaptive Systems. The appendix to this chapter addresses the managerial implication of reciprocal altruism[1]. In Chapter

1 *Note that Chapter 5 will expand more on the theoretical background to reciprocal altruism and related phenomena in co-evolution.*

4 Hermann Kühnle adds to the complex systems view by adding self-criticality and similarity as conditions for autonomous agents in networks. Rob Dekkers offers an outlook on how to combine this view with co-evolutionary models, game theoretical approaches and network theories (Chapter 5). During the years to come, we might expect that further elaboration of the complex systems view in its widest sense will add to our understanding of the behaviour of agents in industrial networks and to the improvement of coordination mechanisms between loosely connected entities.

Part II: Control and Coordination in Industrial Networks

The two contributions in Part II elaborate on the coordination and control processes in industrial networks. In Chapter 6 Petri Helo *et al.* describe an open source solution for managing the logistics between firms in an industrial network with loosely connected entities. Their approach supports the sustainability of Dispersed Manufacturing Networks, through the possibility to connect firms without (severe) interventions in internal systems and processes. In Chapter 7, Hossein Sharifi *et al.* present an integrated approach to facilitate dynamic and simultaneous design and development of products and supply chains contributing to the agility of networks. This should enhance existing practices and approaches to the product development process as well as supply chain development and management. These contributions offer solutions to improve and to facilitate improved coordination and control in industrial networks with interdependent agents.

Part III: International Issues of Industrial Networks

The final contributions in Part III touch on the international aspects of industrial networks. Joachim Kuhn expands on the evolution of the automotive industry when creating dispersed manufacturing networks, to reach local markets and to battle competitive pressures by reducing cost (Chapter 8). Stephen Smith *et al.* build on this proposition by investigating to what extent companies have geographically dispersed their manufacturing capacity (Chapter 9). The contribution by Mahendrawathi Er and Bart MacCarthy in Chapter 10 to this book shows these effects of globalisation on supply chains by the simulation model they have developed. Finally, Harsh Karandikar in Chapter 11 offers an unique model for managing the transition of global engineering networks, moving away from engineering management based on the monolithic company. These contributions align with the need for improving decision-making at managerial level while proving that operational issues will determine the success of managing dispersed manufacturing or for smaller firms the participation in loosely connected networks spanning the globe.

1.4 Concluding Remarks

All these contributions shed some light on how we might view networks of loosely connected entities, particularly Dispersed Manufacturing Networks. At the same time as generating answers to the challenges for these networks, they also create new questions that need new avenues for research. In that sense, the call of Camarinha-Matos and Afsarmanesh (2005) seems justified: this is a new discipline!

References

Barney, J., Wright, M. and Ketchen Jr., D.J. (2001) "The resource-based view of the firm: Ten years after 1991", Journal of Management, Vol. 27, No. 6, pp. 625–641.

Barney, J.B. (1991) "Firm resources and sustained competitive advantage", Journal of Management, Vol. 17, No. 1, pp. 99–120.

Bennett, D. and Dekkers, R. (2005) "Industrial Networks of the Future - A Critical Commentary on Research and Practice", in Proceedings of the 12th International EurOMA Conference, Budapest, 19-22 June.

Biggiero, L. (1999) "Market, hierarchies, networks, districts: A cybernetic approach", Human Systems Management, Vol. 18, No. 2, pp. 71–86.

Brasseul, J. (1998) "Une Revue des Interprétations de la Révolution Industrielle", Revue Région et Développement, No. 7, pp. 1–74.

Camarinha-Matos, L.M. and Afsarmanesh, H. (2005) "Collaborative Networks: a new scientific discipline", Journal of Intelligent Manufacturing, Vol. 16, No. 4–5, pp. 439–452.

Chakravarty, A.K. (2001) Market driven enterprise: products, supply chains, and manufacturing, Wiley, New York.

Das, T.K. and Teng, B.-S. (2001) "Trust, Control, and Risk in Strategic Alliances: An Integrated Framework", Organization Studies, Vol. 22, No. 2, pp. 251–283.

Dekkers, R. (2005) (R)Evolution, Organizations and the Dynamics of the Environment, Springer, New York.

Dekkers, R. and Luttervelt, C.A. van (2006) "Industrial networks: capturing changeability?" International Journal of Networking and Virtual Organisations, Vol. 3, No. 1, pp. 1–24.

Dyer, J.H. and Nobeoka, K. (2000) "Creating and Managing a High-Performance Knowlede-Sharing Network: The Toyota Case", Strategic Management Journal, Vol. 21, No. 3, pp. 345–367.

Dyer, J.H. and Singh, H. (1998) "The Relational View: Cooperative Strategy and Sources of Interorganizational Competitive Advantage", Academy of Management Review, Vol. 23, No. 4, pp. 660–679.

Freiling, J. (1998) "Kompetenzorientierte Strategische Allianzen", io Management, Vol. 67, No. 6, pp. 23–9.

Grant, R.M. (1996) "Toward A Knowledge-Based Theory Of The Firm", Strategic Management Journal, Vol. 17, Special Issue, pp. 109–122.

Greif, A. (1996) "Economic History and Game Theory: a Survey", in: R.J. Aumann and S. Hart (Eds.) Handbook of Game Theory, North Holland, Amsterdam, pp. 1989–2024.

Gulati, R. (1998) "Alliances and Networks", Strategic Management Journal, Vol. 19, No. 4, pp. 293–317.

Gulati, R., Lawrence, P.R. and Puranam, P. (2005) "Adaptation in Vertical Relationships", Strategic Management Journal, Vol. 26, No. 5, pp. 415–440.

Gulati, R., Nohria, N. and Zaheer, A. (2000) "Strategic Networks", Strategic Management Journal, Vol. 21, No. 3, pp. 203–215.

Hamel, G. and Prahalad, C.K. (1994) Competing for the Future, Harvard Business School Press, Boston, MA.

Hardy, C., Phillips, N. and Lawrence, T.B. (2003) "Resources, Knowledge and Influence: The Organizational Effects of Interorganizational Collaboration", Journal of Management Studies, Vol. 40, No. 2, pp. 321–347.

Hemphill, T.A. and Vonortas, N., S. (2003) "Strategic Research Partnerships: A Managerial Perspective", Technology Analysis & Strategic Management, Vol. 15, No. 2, pp. 255–271.

Hoopes, D.G., Madsen, T.L. and Walker, G. (2003) "Why is there a Resource-Based View? Toward a Theory of Competitive Heterogeneity", Strategic Management Journal, Vol. 24, No. 10, pp. 889–902.

Huemer, L. (2004) "Balancing between stability and variety: Identity and trust trade-offs in networks", Industrial Marketing Management, Vol. 33, No. 3, pp. 251–259.

Jarillo, J.C. (1988) "On Strategic Networks", Strategic Management Journal, Vol. 9, No. 1, pp. 31–41.

Karlsson, C. (2003) "The development of industrial networks: Challenges to operations management in an extraprise", International Journal of Operations & Production Management, Vol. 23, No. 1, pp. 44–61.

Kaufman, A., Wood, C.H. and Theyel, G. (2000) "Collaboration and Technology Linkages: a Strategic Supplier Typology", Strategic Management Journal, Vol. 21, No. 6, pp. 649–663.

Kogut, B. (1989) "The stability of joint ventures: Reciprocity and competitive rivalry", The Journal of Industrial Economics, Vol. 38, No. 2, pp. 183–198.

Kogut, B. (2000) "The Network as Knowledge: Generative Rules and the Emergence of Structure", Strategic Management Journal, Vol. 21, No. 3, pp. 405–425.

Lee, W.B. and Lau, H.C.W. (1999) "Factory on demand: the shaping of an agile production network", International Journal of Agile Management Systems, Vol. 1, No. 2, pp. 83–87.

Levin, B.M. (1998) "Strategic networks: The emerging business organization and its impact on production costs", International Journal of Production Economics, Vol. 56–57, pp. 397–405.

Medcof, J.W. (2001) "Resource-based Strategy and Managerial Power in Networks of Internationally Dispersed Technology Units", Strategic Management Journal, Vol. 22, No. 11, pp. 999–1012.

Miles, R.E. and Snow, C.C. (1984) "Fit, Failure and the Hall of Fame", California Management Review, Vol. 26, No. 3, pp. 11–28.

Nassimbeni, G. (1998) "Network structures and co-ordination mechanisms: A taxanomy", International Journal of Operations & Production Management, Vol. 18, No. 6, pp. 538–554.

New, S. and Mitropoulos, I. (1995) "Strategic networks: morphology, epistomology and praxis", International Journal of Operations & Production Management, Vol. 15, No. 11, pp. 53–61.

O'Neill, H. and Sackett, P. (1994) "The Extended Manufacturing Enterprise Paradigm", Management Decision, Vol. 32, No. 8, pp. 42–49.

Robertson, P.L. and Langlois, R.N. (1995) "Innovation, Networks, and Virtual Integration", Research Policy, Vol. 24, No. 4, pp. 543–562.

Thorelli, H.B. (1986) "Networks: Between Markets and Hierarchies", Strategic Management Journal, Vol. 7, No. 1, pp. 37–51.

Uzzi, B. (1997) "Social Structure and Competition in Interfirm Networks: The Paradox of Embeddedness", Administrative Science Quarterly, Vol. 42, No. 1, pp. 35–67.

Vangen, S. and Huxham, C. (2000) "Building Trust in Inter-organizational Collaboration", in Proceedings of the Academy of Management Research Conference, Toronto, August 4–9.

Wernerfelt, B. (1984) "A Resource-Based View of the Firm", Strategic Management Journal, Vol. 5, No. 2, pp. 171–180.

2 Industrial Networks of the Future: Review of Research and Practice

Rob Dekkers
University of the West of Scotland

David Bennett
Aston University

Abstract Academic researchers have followed closely the interest of companies in establishing industrial networks by studying aspects such as social interaction and contractual relationships. But what patterns underlie the emergence of industrial networks and what support should research provide for practitioners? Firstly, it appears that manufacturing is becoming a commodity rather than a unique capability, which accounts especially for low-technology approaches in downstream parts of the network, for example in assembly operations. Secondly, the increased tendency towards specialization has forced other, upstream, parts of industrial networks to introduce advanced manufacturing technologies to supply niche markets. Thirdly, the capital market for investments in capacity, and the trade in manufacturing as a commodity, dominates resource allocation to a larger extent than previously was the case. Fourthly, there is a continuous move towards more loosely connected entities that comprise manufacturing networks. More traditional concepts, such as the "keiretsu" and "chaibol" networks of some Asian economies, do not sufficiently support the demands now being placed on networks. Research should address these four fundamental challenges to prepare for the industrial networks of 2020 and beyond.

Keywords Agile manufacturing, Core competencies, Global operations, International issues, Keiretsu networks, Manufacturing strategy, Outsourcing

2.1 Introduction

In recent years, practitioners and researchers have started to look increasingly at companies as part of networks within which they operate. This is often because of the possibilities offered by information technology and data-communication, globalization of markets and the increasing tendency of companies to specialize (e.g. Karlsson, 2003). These possibilities give firms easier access to the capabilities and resources of others, moving them further away from the traditional logic behind the make-or-buy decision; even though this particular manufacturing decision still attracts attention from researchers to develop appropriate models (e.g. Cáñez *et al.*, 2000; Humphreys *et al.*, 2002; Probert, 1997). Additionally, the world of management has seen an abundance of theories that might have been adequate to deal with the contemporary challenges for some enterprises, but not for many others

(Fischer and Hafen, 1997; Micklethwait and Wooldridge, 1996). The notion of core competencies and the concept of Lean Production serve as examples of such theories that address questions relating to Supply Chain Management in the context of industrial networks; but it could be questioned whether they really deal with the characteristics of networked organizations. Capello (1996, p. 496) supports this statement by noting that insufficient is known about the failing of networks. In this chapter, the authors argue that industrial networks require the adaptation of existing theories to fit their particular characteristics as wells as the development of grounded theories based on the unique characteristics of industrial collaboration.

2.1.1 Brief History of Industrial Networks

Although the study of industrial networks has attracted recent attention among researchers, there was already an awareness of the implications caused by the

Table 2.1. *Evolution of organizational forms (Miles and Snow, 1984, p. 19). The table indicates the evolution of organizational forms that are both internally and externally consistent. Miles and Snow state in their paper that a minimal fit is necessary for survival, a tight fit associates with corporate excellence, and an early fit provides a competitive advantage. Therefore, dynamic networks (industrial networks) require both internal fits and external fits, giving early adopters a competitive advantage.*

Period	Product-market strategy	Structure	Inventor or early user	Core activity and control mechanisms
1800 -	Single product or service. Local/ regional markets.	Agency	Numerous small owner-managed firms.	Personal direction and control.
1850 -	Limited, standardized product or service line. Regional/ national markets.	Functional	Carnegie Steel.	Central plan and budgets.
1900 -	Diversified, changing product or service line. National/ international markets.	Divisional	General Motors, Sears, Roebuck, Hewlett-Packard.	Corporate policies and division profit centres.
1950 -	Standard and innovative products or services. Stable and changing markets.	Matrix	Several aerospace and electronic firms.	Temporary teams and lateral resource allocation devices such as internal markets, joint planning systems, etc.
2000 -	Product or service design. Global, changing markets.	Dynamic network	International/ construction firms. Global consumer goods companies. Selected electronic and computer firms (e.g. IBM).	Broker-assembled temporary structures with shared information systems as basis for trust and co-ordination.

particular characteristics of networked organizations (Håkansson and Johanson, 1988; Wiendahl and Lutz, 2002). In particular, academic interest has centred on two periods in the past. The first of these was during the 1970s and 1980s when attention was focused on Japanese manufacturing concepts and techniques, including Just-In-Time (JIT), co-production and "keiretsu" networks. The second period started during the 1990s, after the bursting of Japan's "bubble" economy, as a consequence of the drive for even lower cost, greater efficiency, and responsiveness to customer demands. This resulted in a more formal recognition of the networked organization as a follow-up to the paradigm of core competencies and the consequent escalation in outsourcing. The earlier overview by Miles and Snow (1984) illustrated the move from the simpler paradigms to more complicated forms of network-based organizations that have subsequently been witnessed in recent years (see Table 2.1) and consequently have attracted academic deliberation.

The establishment and emergence of industrial networks is closely related to the subject of manufacturing strategy. Since Skinner's seminal work in this area (Skinner, 1969), manufacturing has been recognized as a fundamental cornerstone for achieving corporate competitive advantage. Although it recognises the traditional and limited perspective of considering low cost and high efficiency as dominant objectives within manufacturing strategy, this earlier work of Skinner is still rooted in the tradition that economies of scale provide competitive opportunities (Penrose, 1963, pp. 260–265). That tradition has given rise to the monolithic company driven by forward and backward integration (Chandler, 1977), which implied an emphasis on the coordination of operations. Only later does Skinner consider the role of smaller-scale units that may now be regarded as elements of an industrial network (Skinner, 1974), while subsequently questioning the traditional efforts towards productivity improvement through making large capital investments in manufacturing (Skinner, 1986). According to Sturgeon (2002, pp. 8–10), American firms – compared with most Asian and many European companies – have generally placed manufacturing in a low position on the hierarchy of corporate esteem. However, in contrast to Sturgeon's belief, it is argued here that this is also the case for European firms. For example, most companies still regard efficiency as the main objective of their production departments in a survey amongst Spanish companies (Avella, 1999). Consequently, during the 1960s and 1970s the make-or-buy decision was at the heart of Operations Management research. Then, in the 1980s, the interest in Japanese manufacturing techniques, including partnerships with suppliers, sparked the next step towards models for collaboration and Supply Chain Management using JIT principles, while in the early 1990s the concept of core competencies led to renewed interest in outsourcing models. Later the "over-the-wall" tactics of outsourcing made companies examine the networks they had created while managing these from a traditional cost perspective (Dekkers *et al.*, 2002). In the end, the increasing attention paid to networks has not challenged the proposition of Skinner that manufacturing is of paramount importance to industrial performance; and it has not altered that the most common view of manufacturing (including manufacturing networks) is taken from the traditional cost perspective.

2.1.2 The Impact of Globalization

The awareness that has been created that manufacturing strategy comprises more than cost-driven objectives, e.g. meeting customer demands, has created a wider array of perspectives for manufacturing; these perspectives on manufacturing strategy, complemented by the influence of advances in information and communication technology together with globalization and specialization, foster the specific characteristics of industrial networks, i.e. collaboration, decentralization and inter-organizational integration (O'Neill and Sackett, 1994). In these three fields each change in itself requires adaptations by companies and the influence of several of these shifts leverage the need for adequate responses. For example, collaboration requires not only solutions in advanced software, it should also account for the management of industrial networks in an international context whereby individual companies set their own course and develop over time (decentralization). Conversely, efficient international collaboration depends on the appropriate deployment of information and communication technology. The intricate interdependencies of these characteristics transform industrial networks into collaborative efforts that have a large number and wide variety of resources at their disposal especially to meet a greater range of customer demands.

This has caused a change in the prevailing attitude towards resource allocation due to the emergence of the industrial network paradigm. The need for proximity of supply, following the theories about co-production, has required a strong interaction between customers and suppliers. Consequently much research has focused on the need for economic clusters (e.g. Porter, 1990). There are examples of these tendencies changing, like Daimler Chrysler's announcement in 2000 that suppliers need to deliver in six days (rather than 1–2 days previously, with close geographical proximity). It illustrates the different views towards supplier selection and purchasing management that are emerging now; the different views enable to the supplier base to be considered as a network rather than a set of individual actors.

Not only has the scene for suppliers to any industry changed, but many more countries have also followed an active path towards developing relevant economic and industrial competencies, reinforcing the establishment of supply networks. For example, the Thai government has deliberately set out to strengthen its automotive sector by attracting foreign companies in that industry (Katayama, 1999). By contrast, during the 1990s, MIT undertook a study that led to a warning about the decline of manufacturing industry in the U.S.A. (Lester, 2003). However, more recently the U.S.A. has adopted a more progressive approach with the study on Visionary Manufacturing Challenges (NRC, 1998), the U.K. government stimulated the creation of Innovative Manufacturing Research Centres (Dekkers and Wood, 2007) and for the first time the Dutch government set out a research strategy to support the manufacturing industry (de Vaan *et al.*, 2002). Consequently a complex pattern has emerged with the industrial base undergoing shifts by moving to developing countries, emerging countries entering the manufacturing arena, and a revival of some traditional industrialized countries, thus making the situation more dynamic than ever before; in the end, these national policies have only encouraged more extended industrial networks.

At the same time, the make-up of industrial networks has undergone changes, too. The external drivers (such as the move from make-or-buy to co-production or alliances and the drive for flexibility of manufacturing) as well as the internally oriented concepts (such as the attempts to apply Computer Integrated Manufacturing and the use of production cells) demonstrate a continuous move towards more loosely connected industrial entities for manufacturing. The requirement for greater flexibility also impacts on the trend to increase the amount of customization and production of goods on-demand (Lee and Lau, 1999). Contemporary changes in industries point to a further repositioning along the dimension of loosely connected entities, with increasing pressure to respond to market opportunities and to increase flexibility.

2.1.3 Scope of Chapter

Following the moves made by companies as identified previously, this chapter explores the concept of industrial networks for manufacturing. It aims to visualize an approach for industrial networks of the future, i.e. for the next 15 years and beyond, based on ongoing research and additional considerations. Firms are operating increasingly as parts of industrial networks, although the situation is extremely fluid and the stage has not yet been reached where networks are configured optimally and network operations have reached a stage of maturity. Consequently, the chapter also aims at contributing to the research agenda and making a contribution to foundations for generating grounded theory about industrial networks.

Initially, in Section 2.2, this chapter examines the types of traditional networks that have been identified, together with the reasons for which they have been formed and their advantages and weaknesses. This includes a critique of the traditional "keiretsu" and "chaibol" networks based on conglomerate structures that formed the basis of Japan's and Korea's economic success. In Section 2.3, this chapter addresses how future networks will be shaped by discussing three contributory and related topics: network configuration, manufacturing as commodity and added value within networks. The chapter then moves to present the outlines of a research agenda in Section 2.4. This contribution to directing research into industrial networks uses a blend of illustrations (from the business literature), findings of previous studies by others together with results from research by the authors, to construct a picture of how future networks might look and behave.

2.2 Traditional Views About Networks

The study of networks as a key aspect of industrial organization goes back to the 1980s with the seminal work of Håkansson at Uppsala University, who defined networks as sets of more or less specialized, interdependent actors involved in exchange processes (Håkansson and Johanson, 1988, 1992). Around the same time the study of urban, networked organizations in the industrialized regions of northern Italy recognized the importance of networks for improving logistical efficiency (Camagni, 1988, 1993). The participation of companies in these networks depends on managing product development, both at the level of the network and the individual companies, and on managing manufacturing processes. Within the overall primary process of most companies the connection between product development and manufacturing strategy has yet to result in conceptual approaches to establish this vital link. Conducting a study into sequential and simultaneous approaches to engineering new products, Riedel and Pawar (1998) highlight that the concepts of design and manufacturing are not linked in the literature and that the interaction of product design and manufacturing strategy is under-researched. Spring and Dalrymple (2000) came to a similar conclusion when examining two cases of product customization, where manufacturing issues received little attention during design and engineering. The only concept that addresses these issues so far is the one of the Order Entry Points (more commonly known as Order Decoupling Points; see Figure 2.1). Order Entry Points and modular product architecture typically concern the optimization of Make-to-Order production concepts and might include product development and engineering activities (Dekkers, 2006a). Introducing a different perspective, Smulder *et al.* (2002) proposed a typology of intra-firm and inter-firm interfaces, therewith also connecting product development and production; yet, this typology has still to be adopted in practice. Henceforth, the emerging paradigm of industrial networks, if it is to be successful, should address this matter of creating a link between manufacturing strategy and product development.

But do we find this link in current concepts for industrial networks? Three mainstream Operations Management concepts in this area dominate thinking about the industrial network paradigm: core competencies, agile manufacturing, and "keiretsu" and "chaibol" arrangements. As shown in the next three subsections, these concepts focus mainly on issues of manufacturing and less on product development, except in general terms.

2.2.1 Core Competencies and Outsourcing

According to Friedrich (2000), focusing on core competencies (Prahalad and Hamel, 1990) and outsourcing (Gilley and Rasheed, 2000) raises the key issue of which areas of production are needed to maintain the value chain and on which areas the company should concentrate for achieving optimal performance. Prahalad and Hamel subtly expand the view of technology from a broadly described concept, the importance of which is determined by its support of the corporate mission, to a specific source of corporate uniqueness. In Prahalad and Hamel's view core competencies represent the collective learning of the organization, especially concerning how to coordinate diverse production skills and integrate multiple streams of technology. However, the application of this theory does not lead directly to a clearly defined strategy for global manufacturing or manufacturing networks. Only when core competencies are linked to decision-making will a manufacturing strategy be found that offers guidelines on decision-making for resource acquisition and capacity management (Hayes and Pisano, 1994).

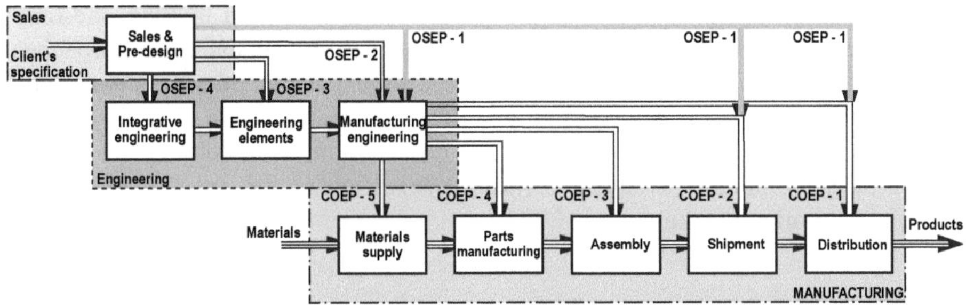

Figure 2.1. *Position of the Order Entry Points in the primary process of design, engineering, manufacturing and logistics. To simplify the figure, points of stock (inventory) have been omitted. OSEP-1 (Order Specification Entry Point) indicates that customer requirements are directly transferred into production instructions, while OSEP-4 points to Engineering-to-Order. Similarly in the material flow: COEP-1 (Customer Order Entry Point) tells that orders are delivered directly from stock, while COEP-5 marks Make-to-Order.*

Given the (often unquestioned) popularity of the concept of core competencies and its implications, how does industry manage the increasing scope of outsourcing? A study by Dekkers (2005a) based on 6 case studies (four in the Netherlands, one in China and one in Indonesia) points to poor control of outsourcing by industrial companies. Most of the case companies, with primary processes based mainly on Engineering-to-Order and Make-to-Order, experienced problems with implementing manufacturing strategies. Ideally, the manufacturing strategy of these companies should address their core competencies and opportunities for outsourcing. All the case companies, except one, had done so, implicitly or explicitly; but mostly this strategy had not been transferred to guidelines for implementation, which is why decision-making occurred at random or through opportunity. There was no feedback about suppliers' performance to the stages of design and engineering, so sometimes problems would recur regularly. None of the companies followed an active approach towards supplier networks for the purpose of expanding their technological capabilities. Operational control posed additional challenges, although not all companies were aware of the impact this caused. In two cases the in-house production of some manufacturing processes proved more beneficial than outsourcing, although this was only discovered with hindsight. All the companies reported problems with on-time deliveries by suppliers, with some of these problems arising from reactive interventions rather than pro-active securing of purchase orders. In summarizing these case results, it can be concluded that operational control in these companies created a wide variety of problems. That is evidenced by poor operational control and poor integration between design, engineering, purchasing and manufacturing; additionally, it indicates that the simplified view of core competencies and outsourcing might have strong limitations.

Still today, even though insight into effective manufacturing strategies has progressed, many approaches for outsourcing rely on the deployment of criteria derived from traditional make-or-buy decisions. However, the rise of industrial networks creates the need for frameworks that take account of early supplier involvement, collaboration, and inter-organizational integration. Also, decision-making concerning the allocation of resources has shifted from making one-time decisions to continuous evaluation and reallocation. Current outsourcing approaches rarely account for this, and, hence there is a need for expansion of criteria to include those suitable for networks. Current practices for management and control of outsourcing still focus largely on minimizing costs and meeting delivery schedules, while research into outsourcing has not yet investigated the specific impact of industrial networks (Dekkers *et al.*, 2002).

2.2.2 Agile Manufacturing Networks

In contrast to the concept of outsourcing, the approach of agile manufacturing relies more strongly on the exploitation of loosely connected networks than earlier concepts such as Lean Production (Nagel *et al.*, 1991; Burgess, 1994; Katayama and Bennett, 1999). Co-production (and subsequently Lean Production) had already introduced a higher degree of outsourcing and improved control through Supply Chain Management, although here the networks used were more closely connected keiretsu or chaibol types involving cross-ownership, as described later. In contrast to the internal focus of Lean Production, the paradigm of agile manufacturing has an external focus and is primarily concerned with the ability of enterprises to cope with unexpected changes, to survive against unprecedented threats from the business environment, and to take advantage of changes as opportunities (Goldman *et al.*, 1995). Similarly, Kidd (1995) recognizes two main factors within the concept of agility, i.e. responding to changes in appropriate ways and in due time, and taking advantage of the opportunities resulting from change. This means that an agile manufacturing enterprise marshals the best possible resources to provide innovative (and often customized) products, with the flexibility to adjust the product and offer rapid delivery, and with the high level of efficiency required to be competitive and profitable (Goldman and Nagel, 1993, p. 19). The concept of agile manufacturing stresses two interconnected main processes:

i) the development of innovative products;
ii) the manufacturing and distribution of products.

These two processes should meet lead-time requirements (time-to-market, time-to-volume and delivery time) and flexibility requirements (to meet market opportunities and respond to market demands) (Stock *et al.*, 1999). A reconfigurable structure becomes a prerequisite for optimizing the capabilities of an organization for each business opportunity (Ross, 1994), which itself requires more loosely connected entities.

However, even the new types of agile manufacturing networks are often not designed within an international context and may still be suboptimal where acquisitions have taken place resulting in an inherited supplier base. Therefore, the notion of building international manufacturing networks is now a prevalent concern where competitiveness derives from an ability to garner and integrate resources from a number of different geographical sources. The basic principles for building a manufacturing network have been described by Mraz (1997), who identifies four categories of resources (i.e. players) that can be used within the network: industrial design consultants, product development consultants, contract manufacturers, and Original Equipment Manufacturers (OEMs). These last two players also demonstrate the options available for the production of complex products and their relative advantages and disadvantages, with the contract manufacturing approach typically involving external industrial design and product development, and the OEM approach typically retaining these activities in-house. A hybrid of these two forms can be found in the case of the Brazilian aircraft manufacturer Embraer (Empresa Brasiliera de Aeronáutica SA), which, with its network of risk sharing partners, was able to greatly accelerate the development and launch of the ERJ-170/190 series of regional jets. Hence, adequate suppliers' bases, with possibly an international dimension, reinforce the performance during product development (reduced time-to-market) and manufacturing (improved performance to deliver) to the advantage of OEMs and their supplier networks.

The international dimension to designing agile manufacturing networks is also considered by Lee and Lau (1999), who use the example of firms in Hong Kong and the Pearl River Delta to provide a "Factory-on-Demand"-concept within the context of manufacturing networks. Shi and Gregory (1998) have contributed by proposing the mapping of configurations for international manufacturing networks as a means of providing support for decision-making. Presentations by companies at the *9th Annual Cambridge International Manufacturing Symposium* in 2004, organized by the University of Cambridge, have shown that there are two strategic directions for international manufacturing networks: "rationalization" (with manufacturing units, sometimes including product development, specializing on product ranges) and "globalization" (taking the opportunity to outsource operations or establish alliances). As frequently evidenced in the literature (e.g. Shi, 2003), the current drive for globalization by companies places its emphasis more on optimization within existing conditions and less on capturing new market opportunities, even for the opportunities these international manufacturing networks offer.

2.2.3 Keiretsu and Chaibol Networks

Unlike the networks of Western companies, the "keiretsu" and "chaibol" networks that formed the basis of Japan's and Korea's economic success were based on conglomerate structures. However, more recently these structures have proved less capable of meeting the need for speed of change, flexibility, and cost reduction that have been the key aspects of industrial management following the Asian economic crisis of the late 1990s (Business Week, 1999). At the same time, organizations that attempted to replicate the keiretsu concept outside Japan have encountered severe problems, making them rethink their plans to create similar supply networks (Stein, 2002).

A major weakness of the traditional keiretsu and chaibol networks has been their domestic focus and cross-ownership between companies in the network. This has hindered how they can respond effectively to the globalization of manufacturing (Bennett, 2002). It has also created difficulties as end-product manufacturers have moved offshore and taken them beyond the reach of domestically based network members. Also, the burden of debt resulting from borrowing to support cross-ownership has restricted their ability to develop and fully support international operations. As a consequence of this situation Renault, on taking a controlling interest in Nissan, sought to dismantle its keiretsu supplier network by selling off most of its financial stakes in almost 1,400 companies (Zachary, 2001). This indicates that companies deploying traditional networks are searching for different concepts to manage their suppliers.

However, despite these concerns, a study by McGuire and Dow (2003) still shows that throughout the first half of the 1990s the keiretsu system remained strongly in place. At the same time, they conclude that the continued move toward globalization of capital markets in Japan and ongoing regulatory change may potentially impact networking and performance implications. Apart from the problems that can arise when there is cross-ownership between companies, the main criticism of the keiretsu relates to its lack of flexibility and responsiveness. The answer to this criticism has therefore been to propose the creation of agile networks (Tian *et al.*, 2002).

2.2.4 Traditional Views on the Wane

Despite the theoretical ability of agile manufacturing to provide greater flexibility and responsiveness there are still questions about whether it can address the characteristics of networks, i.e. decentralization of decision-making, inter-organizational integration and collaboration. The Special Issue on Dispersed Manufacturing Networks underlines the fact that progress is being made slowly (Dekkers, 2006b). The questions around the paradigm for networks that consist of loosely connected entities only demonstrate that we still know little about their behaviour. Nevertheless, many developments in information technology and data-communication allow interfacing in networked manufacturing; for example, as Boeing has done for the 787 Dreamliner. The current problems with production can be traced back to selection processes of suppliers (even supported by sophisticated software applications that failed to solve the process of interaction). The lack of synchronization between the possibilities of information technology and the limited understanding of the actual behaviour of entities (or agents for that matter) have only increased instability in relationships, giving greater cause for issues of trust and power fueling uncertainty and opportunistic behaviour. At the same time, interrelationships have become more demanding and have limited the capabilities of parties to operate within each other's constraints. Industrial companies demand partnerships, but these sometimes appear to be forcibly driven by a one-sided strategy rather than being based on a true bilateral relationship. With the reduced capability to maintain long-term relationships, partners in industrial networks need different ways of interacting, sometimes facilitated by applications in information technology and data-communication (extending to both the domain of manufacturing and the domain of product development and engineering).

2.3 Future Networks

There are now many emerging possibilities offered by information technology and data-communication methods. Some of these include planning methodologies (Frederix, 2001), Smart Supply Chains (Noori and Lee, 2002), globalization of markets (Karlsson, 2003) and the ongoing specialization of firms. They drive companies to concentrate on competencies and, consequently, enable them to move from centralized, vertically integrated, single-site manufacturing facilities to geographically dispersed networks of resources (Stock *et al.*, 1999). These simultaneous developments foster the specific characteristics of (international) networks, which require adaptations by companies to fit these characteristics.

2.3.1 Network Configuration

The dilemma with these networks extends to the problem of achieving a balance between having independent agents and controlling processes to meet performance, which requires a strong interaction between these agents. Virtual organizations, which can be considered as a further manifestation of networks, might display instability between the model of pure outsourcing and the establishment of more traditional alliances (Roosendaal, 2000). Even alliances, which are perceived as more stable relationships between firms, usually dissolve over time or result in mergers (Kogut, 1989). The network is optimized locally and creates power shifts if the balance moves towards independence of agents, depending on the uniqueness of their resources (Medcof, 2001). Also, flexibility might be lost in the short and medium term through the creation of alliances or mergers (Mody, 1993). Therefore, research needs to be undertaken to reveal whether this dilemma of balance between control and change in networks can be resolved.

The principal characteristic of industrial networks is their ability to capture market opportunities and to adapt to changes in the environment. Collaboration with other companies has a significant impact on the capabilities of a network. Hitherto, the dynamic capability has equated to "changeability", which Milberg and Dürrschmidt (2002) define as the sum of (i) flexibility, defined as the capability to operate in a wider space on certain dimensions of business management, and (ii) responsiveness, defined as the ability to handle emerging changes in the environment. Thus changeability is a measure of the total changes the environment demands of an organization or network (Wiendahl and Lutz, 2002). Sometimes, the sacrifices in a given production system to obtain flexibility (i.e. capturing market opportunities and adapting) exceed the derived benefits.

Each market opportunity requires an adequate response from an industrial network. The flexibility of a network relies on the deployment of resources to capture these market opportunities and thereby needs a control structure and organizational structure that fits the actual demand. Theory about organizational design distinguishes the process structure, the control structure, the organelle structure, and the hierarchy (Dekkers, 2000); the organelle structure is based on the grouping of (business) process or activitites to address performance requirements.

The methodology for the design of organizations assumes a linear process when designing each of these structures consecutively, see Figure 2.2, even though this process should be considered iterative. In this approach the design of the organelle structure is key to meeting performance demands by customers; which leads Dekkers and van Luttervelt (2006, p. 13) to propose a model for reconfiguration of networks, see Figure 2.3. Industrial networks provide the opportunity for optimizing each of the four structures to take place independently and that through the connections between these structures, as present in the value chain and as individual agents, optimization will occur over time.

Another phenomenon is the increasing participation of Small and Medium-sized Enterprises (SMEs) in international manufacturing networks (Tesar *et al.*, 2003), which has been enabled through the factors identified by Lall (2000) as contributing to the increase in SME competitiveness. Bennett and Ozdenli (2004) have studied the role of several SMEs in international manufacturing networks. The SMEs were based in industrialized countries, developing countries and transition economies. The analysis of the cases shows that they are motivated largely by the desire to extend their reach and a wish to begin establishing a global presence. It also shows that control and commitment are two major determinants for SMEs and international manufacturing networks, so managers must think carefully about how much control they want to have (or should have) within the network. This concerns the electronic and virtual integration of companies, so calling on totally new models for dealing with networks (Dekkers *et al.*, 2004). These include matchmaking and brokerage through web services (Field and Hoffner, 2003; Molina *et al.*, 2003) and electronic contracts; these will enable companies to move away from the control paradigm for the monolithic company towards management approaches that fit the emergent properties of networks (Angelov and Grefen, 2003; Barata and Camarinha-Matos, 2003). The concept of complex networks with emerging properties strongly relates to the proposed idea of Open Innovation Systems (Chesbrough, 2003; Tidd, 1995);

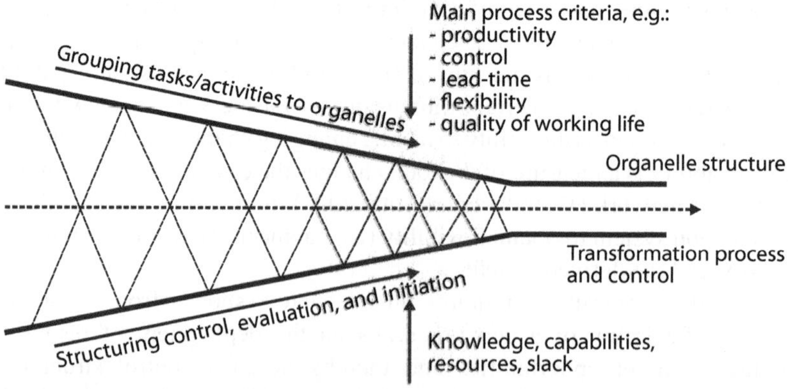

Figure 2.2. *Design process for the organelle structure (Bikker, 1993, pp. 183–188). The organelle structure affects both the grouping of tasks in the primary process as well as the control processes. By subsequent integration and iteration, the design of the organelle structure meets performance requirements.*

the increased interaction between actors in networks requires a rethinking how it happens at all (Dekkers, 2009), whether it concerns manufacturing or innovation and product development.

2.3.2 Manufacturing as a Commodity

An important development influencing the shift in power within manufacturing networks has been the increasing importance of OEMs and, more recently, brand owners (Kotabe and Murray, 2004). Sturgeon (2002) argues that the revival of the American industry during the 1990s can be attributed to what he calls "turnkey" production networks. Essentially, these incorporate the trend towards outsourced manufacturing and an emphasis on branding. To demonstrate this concept, Sturgeon uses the example of the electronics industry, particularly the case of Apple Computer Inc. that contracted SCI Systems for a large part of its manufacturing operations in 1996. A system like a turnkey production network is highly adaptive because it uses turnkey relationships to weave various key production clusters into a global-scale production network based on external economics for OEMs and brand owners.

With the rise in OEMs, especially in the electronics and automotive industries, the concept of outsourcing the production of complete systems and subsystems

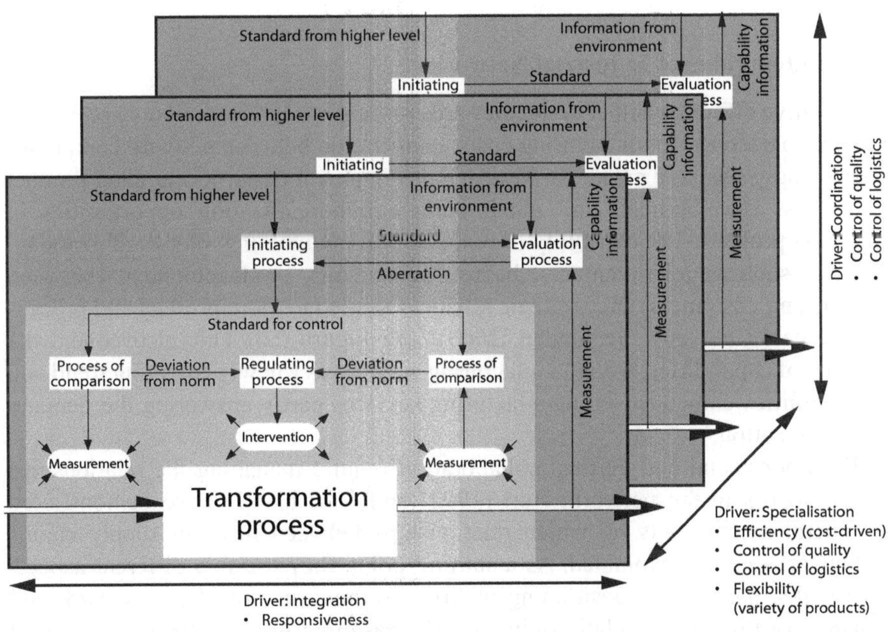

Figure 2.3. *Model for Reconfiguration within networks. Based on different drivers market opportunities call for either integration, specialization or coordination to meet performance requirements. Through predefined organelles for both the primary process and the control processes reconfiguration becomes a preset decision-making process allowing quick responses to changing conditions.*

started to become a common phenomenon. In this way the idea of "tiering" in the supply network was created (Sadd and Bennett, 1999), with power generally reducing towards the lower tiers (with possible exceptions where suppliers are part of much larger companies involved with leading edge technologies). Along with this trend has the idea of manufacturing capacity as a commodity rather than a unique capability for "pushing" products onto growing markets also materialized. At the same time, the focus of technology has also moved upstream with suppliers increasingly turning to advanced manufacturing technologies as a means of competing for orders, while OEMs, especially those based offshore, have tended largely to rely on low-tech assembly techniques for enabling greater agility.

This trend has been taken further under the more recent, and increasingly dominant, regime of brand ownership. A characteristic is the separation of brand from origin of production and the virtually complete transition of manufacturing to a commodity with power residing almost totally with the brand owner; this often causes the brand to be more dominant than the actual product (Joo *et al.*, 2003). In turn, this has led to manufacturing becoming increasingly footloose with international mobility being an important aspect of network design. In particular, this has resulted in the transfer of production capital away from the traditional industrial economies to the low factor cost economies of the Far East and the transition economies of Eastern Europe (Bennett *et al.*, 2001).

2.3.3 Added Value of Industrial Networks

Collaborative efforts, whether or not they are crossing borders, are not only seen as an approach to decrease manufacturing costs; cooperation between network companies is increasingly seen as a means for lowering development costs, accelerating product and process development, and maximizing commercialization opportunities in innovation projects. The capability of building and maintaining inter-organizational networks, such as joint ventures, license agreements, co-development (between suppliers and customers) and strategic alliances has led to more product and process innovations (Ritter and Gemünden, 2003); see Figure 2.4. This also covers the extension of capabilities, with manufacturing services as a newly emerging trend and the capabilities embedded in manufacturing services partly answering the demand for customization.

Both horizontal and vertical collaboration require managing the relationships between actors in the network. Burt (1992) and Uzzi (1997) have demonstrated the general mechanisms by which relationships between firms in supply chains and networks can be explained. As a starting point, they use two different aspects of networks, namely the positioning of firms in the structure of the network and the nature of the mutual relationships. Burt's reasoning implies that the chance of achieving completely radical innovations may decrease if companies establish strong mutual contractual links, such as in supply chains. Links with other companies in the supply chain might be so strong that they prevent a company from successfully implementing an innovation, even if it is in a strategic position to do so. Typically, a successful cooperation strategy consists of three basic elements, i.e. selection of a suitable partner, formulation of clear-cut agreements (getting the project underway), and management of the ongoing relationship. Careful selection of future collaborative partners can prevent many problems and, according to Hagedoorn (1990), the aim should be similarity balanced by complementarity, with similarity referring to the firm's size, resources, and performance. However, of more importance are the required complementarities offered by the cooperation partner, i.e. the combination of complementary activities, knowledge, and skills to realize the desired synergy. The literature on strategic partnerships offers many models to evaluate potential cooperation partners (e.g. Souder and Nassar, 1990). Based on a study of 70 UK based firms in different industry sectors, Bailey *et al.* (1996) even concluded that selecting partners based on their track record in previous collaborations turns out to be a poor basis for future collaboration. These signals indicate that how collaborations can be effectively exploited has not yet been settled.

2.4 Research Agenda for Industrial Networks

The three themes – network configuration, manufacturing as commodity, added value of networks – described in the previous section appear not to be congruent with most of the ongoing research into industrial networks. Nassimbeni (1998) comments that the bulk of available research on networks is devoted to the contractual aspects and social dynamics of inter-organizational relationships, while the dynamic forms of communication and coordination have been neglected, so requiring more attention from researchers. Most likely this originates in the conversion from the hierarchical firm, with direct control of resources and a cross-ownership strategy towards

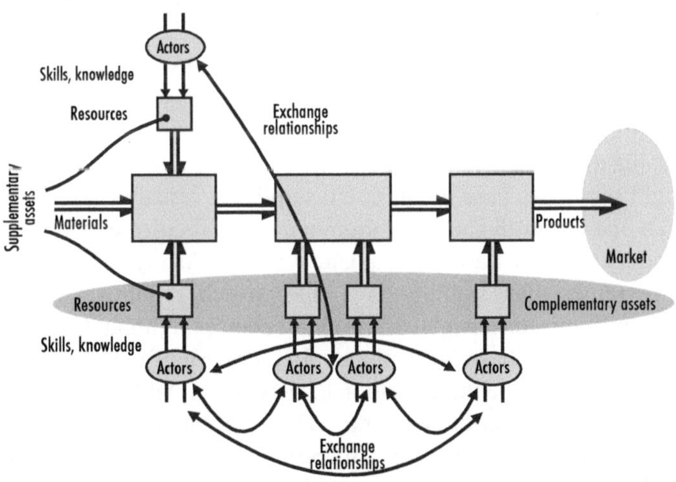

Figure 2.4. *Collaboration Model for the value chain (Dekkers, 2005b, p. 330). Vertical collaboration indicates the capability of actors to manage the supply chain. Horizontal collaboration contributes to the dynamic capability of the network by reallocating resources or creating substitution.*

suppliers, to networks with more loosely connected entities, which is a view also found in Smulder *et al.* (2002).

However, the shift towards more loosely connected entities requires additional theory, models and tools to cope with issues of collaboration, inter-organizational integration and decentralization of decision-making. It is probably more than a decade since the beginnings of academic research into the networked organization (which initially looked at the extended enterprise, etc.). This research has mainly used models from the monolithic company – decision-making on make-or-buy and social dynamics – to further research. Reported findings of research argue that studies should pay more attention to modelling the interaction between agents (Robertson and Langlois, 1995), meaning that a more integrated approach becomes necessary. Therefore, research should consider taking different routes:

- The recent insights in natural sciences and the application of principles of complex systems theory to collaborative enterprise networks as socio-technical systems might yield these complementary approaches. Six themes emerge (Dekkers *et al.*, 2004):
  i. the dynamic description of networks (to respond to market opportunities and shifting demands and to capture the stability of networks themselves);
  ii. coordination possibilities (the networks consist of loosely connected entities, each with their own strategy, and dependent on each other for delivery of products and services);
  iii. radical and integrative innovation (the capturing of new market opportunities and technological prospects, and at the same time taking advantage of individual agent's knowledge and skills);
  iv. path dependency in the evolution of networks (the concepts of evolutionary approaches and concepts like co-evolution and symbiosis applied to industrial networks);
  v. sharing of information across agents (the network as a community of entities that evolve together);
  vi. modelling and representation of industrial networks (to stretch beyond classification and static approaches).

  This might serve as a base for an interdisciplinary research approach.
- Networks operate in dynamic environments and require dynamic approaches, so reflecting Ashby's Law of Requisite Variety (Ashby, 1956). Perhaps even instability rather than stability is a rule, which requires that optimization models should rely on insight from other sciences. Although neural networks incorporate some of these ideas, the explicit criteria of optimization, dispersal, and bifurcation describe the evolution of networks (Dekkers, 2009).
- The efficacy of industrial networks relies on the use of Information and Communication Technology for collaborative engineering, Computer-Aided Production Planning, Supply or Value Chain Management and communication (Maropoulos, 2003; Noori and Lee, 2002), so exceeding the need for logistics integration, which is the main argument of Stock *et al.* (1999). Also, the optimization of structures can be supported by information technology. Helo *et al.* (2006) propose a flexible, integrated web-based logistics management

system for agile supply demand network design, allowing to interface different scheduling agents from different actors. Nevertheless, a lot of development work needs to be done to obtain methodologies, methods and tools to sustain industrial networks as loosely connected entities (Dekkers, 2009).

- The reconfiguration, for which a method should still be developed, allows a more appropriate approach for capturing market opportunities and optimizing performance of networks (see Dekkers and van Luttervelt [2006, p. 19] and Subsection 2.3.1).
- The link between product development and manufacturing needs to be investigated more closely. So far, research has concentrated on Order Entry Points, product families, etc.; but these concepts have limited reach, although they are addressing an important capability of networks: (mass) customization. Particularly, the impact the interface between product development and manufacturing on networks needs attention.

Although the specific research into approaches for networks has progressed, further advances should create insight into optimization and tools to support industrial networks; this is congruent with the remark of Camarinha-Matos and Afsarmanesh (2005, pp. 443-444) that research into Collaborative Networks constitutes a new interdisciplinary domain.

2.5 Conclusion

There is little doubt that the issue of industrial networks has been an important concern to companies needing to compete in the dynamic competitive climate that has demanded greater flexibility, responsiveness and variety as well as responding to pressure on costs. The traditional networks of the past, especially those based on keiretsu or chaibol principles, are less appropriate in today's business conditions and, as a consequence, more loosely connected agile networks have emerged. However, there has been very little research aimed at establishing the patterns that underlie their emergence, and there remains the question of what support such research should provide for practitioners.

This chapter has identified a number of key issues concerning the future of networks, which have been based on a review of the relevant literature and additional considerations. First, network configurations require a control structure and organizational structure that fits actual demand, so companies have started to move away from the control paradigm of the monolithic company towards managing the emergent properties of networks. Second, with the move towards OEMs as network players there has been a greater tendency for manufacturing to become a commodity, which has accelerated under the regime of brand ownership. Third, the added value of industrial networks includes more product and process innovations and the extension of capabilities with manufacturing services. Finally, a number of different routes that research should take if it is to properly reflect and support industrial networks in the coming decade and beyond have been identified.

References

Angelov, S. and Grefen, P. (2003) "The 4W framework for B2B e-contracting", International Journal of Networking and Virtual Organizations, Vol. 2, No. 1, pp. 78–97.

Ashby, W.R. (1956) An introduction to cybernetics, Wiley, New York.

Avella, L. (1999) "Focal points in manufacturing strategic planning in Spain", International Journal of Operations & Production Management, Vol. 19, No. 12, pp. 1292–1317.

Bailey, W.J., Masson, R. and Raeside, R. (1996) "Choosing Successful Technology Development Partners: A Best Practice Model", in Proceedings of the 6th International Forum on Technology Management, Amsterdam, 15–8 October.

Barata, J. and Camarinha-Matos, L.M. (2003) "Coalitions of manufacturing components for shop floor agility - the CoBASA architecture", International Journal of Networking and Virtual Organizations, Vol. 2, No. 1, pp. 50–77.

Bennett, D. (2002) "Current and Future Challenges for International Manufacturing Networks", in Proceedings of the VIII International Conference on Industrial Engineering and Operations Management, UFRGS/ABEPRO, Porto Alegre, Brazil, 23–25 October.

Bennett, D.J., Liu, X., Parker, D., Steward, H.F. and Vaidya, K.G. (2001) "Technology Transfer to China: A Study of Strategy in 20 EU Industrial Companies", International Journal of Technology Management, Vol. 21, No. 1/2, pp. 151–182.

Bennett, D.J. and Ozdenli, O. (2004) "International Manufacturing Networks and the Role of Small and Medium Enterprises", in Proceedings of the ENEGEP X International Conference on Industrial Engineering and Operations Management and XXIV National Conference on Production Engineering, Florianopolis, Brazil, 3–5 November.

Bikker, H. (1993) Lecture notes wb5413, Delft University of Technology, Delft.

Bollinger, J.G. (1998) Visionary Manufacturing Challenges for 2020, National Academy Press, Washington.

Burgess, T.F. (1994) "Making the Leap to Agility: Defining and Achieving Agile Manufacturing through Business Process Redesign and Business Network Redesign", International Journal of Operations & Production Management, Vol. 14, No. 11, pp. 23–34.

Burt, R.S. (1992) Structural Holes: The Social Structure of Competition, Harvard University Press, Cambridge, MA.

Business Week (1999) "Time to Overhaul Japan's Keiretsu Model", Business Week, No. 3620, p. 114.

Camagni, R. (1988) "Functional integration and locational shifts in the new technology industry", in: P. Aydalot and D. Keeble (Eds.) High-technology Industry and Innovative Environments: The European Experience, Routledge, London, pp. 48–64.

Camagni, R. (1993) "Inter-Firm Industrial Networks: The Costs and Benefits of Cooperative Behaviour", Industry & Innovation, Vol. 1, No. 1, pp. 1–15.

Camarinha-Matos, L.M. and Afsarmanesh, H. (2005) "Collaborative Networks: a new scientific discipline", Journal of Intelligent Manufacturing, Vol. 16, No. 4–5, pp. 439–452.

Cánez, L.E., Platts, K.W. and Probert, D.R. (2000) "Developing a framework for make-or-buy decisions", International Journal of Operations & Production Management, Vol. 20, No. 11, pp. 1313–1330.

Capello, R. (1996) "Industrial Enterprises and Economic Space: the Network Paradigm", European Planning Studies, Vol. 4, No. 4, pp. 485–498.

Chandler, A.D., Jr. (1977) The Visible Hand: The Managerial Revolution in American Business, Harvard University Press, Cambridge, MA.

Chesbrough, H.W. (2003) Open innovation: the new imperative for creating and profiting from technology, Harvard Business School Press, Boston.

Dekkers, R. (2000) "Organelle Structures: Strategic Choice of Performance Management of Manufacturing", in Proceedings of the 1st World Conference on Production and Operations Management, Sevilla, 26–30 August.

Dekkers, R. (2005a) "Outsourcing Revisited", in Proceedings of the 18th International Conference on Production Research, Salerno, 31 July–4 August.

Dekkers, R. (2005b) (R)Evolution, Organizations and the Dynamics of the Environment, Springer, New York.

Dekkers, R. (2006a) "Engineering Management and the Order Entry Point", International Journal of Production Research, Vol. 44, No. 18–19, pp. 4011–4025.

Dekkers, R. (2006b) "Guest editorial", Journal of Manufacturing Technology Management, Vol. 17, No. 8, pp. 1017–1021.

Dekkers, R. (2007). "Distributed Manufacturing as Co-Evolutionary System", in Proceedings of the 19th International Conference on Production Research, Valpairaso, 31 July - 4 Aug.

Dekkers, R. (2009) "Distributed Manufacturing as Co-Evolutionary System", International Journal of Production Research, Vol. 47, No. 8, pp. 2031–2054.

Dekkers, R. and Luttervelt, C.A. van (2006) "Industrial networks: capturing changeability?" International Journal of Networking and Virtual Organisations, Vol. 3, No. 1 pp. 1–24.

Dekkers, R., Sauer, A., Schönung, M. and Schuh, G. (2004) "Collaborations as Complex Systems", in Proceedings of the Designing and Operating Global Manufacturing & Supply Networks, 9th Annual Cambridge International Manufacturing Symposium, IMNet/CIM, Cambridge, 25–26 September.

Dekkers, R., Wang, L. and Zhang, S. (2002) "Management and Control of Outsourcing in Dispersed Manufacturing Networks", in Proceedings of the International Manufacturing Conference China, Xiamen, 11–13 October.

Dekkers, R. and Wood, K.L. (2007) Innovative Manufacturing Programme - Overall Mid-Term Review of the Innovative Manufacturing Research Centres, University of Paisley, Paisley.

Field, S. and Hoffner, Y. (2003) "Web services and matchmaking", International Journal of Networking and Virtual Organizations, Vol. 2, No. 1, pp. 16–32.

Fischer, D. and Hafen, U. (1997) "Immer wieder die gleichen Fehler", io Management Zeitschrift Industrielle Organisation, Vol. 66, No. 6, pp. 38–45.

Frederix, F. (2001) "An extended enterprise planning methodology for the discrete manufacturing industry", European Journal of Operational Research, Vol. 129, No. 2, pp. 317–325.

Friedrich, S.A. (2000) "Was ist "Core" und was ist "Non-Core"?" io Management, Vol. 69, No. 4, pp. 18–23.

Gilley, K.M. and Rasheed, A. (2000) "Making More by Doing Less: An Analysis of Outsourcing and its Effects on Firm Performance", Journal of Management, Vol. 26, No. 4, pp. 763–790.

Goldman, S.L. and Nagel, R.N. (1993) "Management, technology and agility: the emergence of a new era in manufacturing", International Journal of Technology Management, Vol. 8, No. 1/2, pp. 18–8.

Goldman, S.L., Nagel, R.N. and Preiss, K. (1995) Agile competitors and virtual organization; strategy for enriching the customer, Van Nostrand Reinhold, Princeton.

Hagedoorn, J. (1990) "Organizational Modes of Inter-Firm Co-Operation and Technology Transfer", Technovation, Vol. 10, No. 1, pp. 17–30.

Håkansson, H. and Johanson, J. (1988) "Formal and informal co-operation strategies in international industrial networks", in: F. J. Contractor and P. Lorange (Eds.) Cooperative Strategies in International Business, Lexington Books, Lexington, MA, pp. 369–379.

Håkansson, H. and Johanson, J. (1992) "A model of industrial networks", in: B. Axelsson and G. Easton (EEds.) Industrial Networks: A New View of Reality, Routledge & Kegan Paul, London, pp. 28-38.

Hayes, R.H. and Pisano, G.P. (1994) "Beyond World-Class: The New Manufacturing Strategy", Harvard Business Review, Vol. 72, No. 1, pp. 77–85.

Helo, P., Xiao, Y. and Jiao, J.R. (2006) "A web-based logistics management system for agile supply demand network design", Journal of Manufacturing Technology Management, Vol. 17, No. 8, pp. 1058–1077.

Humphreys, P., McIvor, R. and Huang, G. (2002) "An expert system for evaluating the make or buy decision", Computers & Industrial Engineering, Vol. 42, No. 2–4, pp. 567–585.

Joo, M.-S., Nakamoto, K. and Nelson, J.E. (2003) "The Shielding Effects of Brand Image Against Lower Quality Countries-of-Origin in Global Manufacturing", Journal of Business Research, Vol. 56, No. 8, pp. 637–646.

Karlsson, C. (2003) "The development of industrial networks: Challenges to operations management in an extraprise", International Journal of Operations & Production Management, Vol. 23, No. 1, pp. 44-61.

Katayama, H. (1999) "Design of a Global Car Production-Logistics System for a Future ASEAN-China Region", International Journal of Operations & Production Management, Vol. 19, No. 5/6, pp. 582-603.

Katayama, H. and Bennett, D.J. (1999) "Agility, Adaptability and Leanness: A Comparison of Concepts and a Study of Practice", International Journal of Production Economics, Vol. 60–61, No. 1, pp. 43–51.

Kidd, P.T. (1994).Agile manufacturing, forging new frontiers, Addison-Wesley, Wokingham.

Kogut, B. (1989) "The stability of joint ventures: Reciprocity and competitive rivalry", The Journal of Industrial Economics, Vol. 38, No. 2, pp. 183–198.

Kotabe, M. and Murray, J.Y. (2004) "Global Sourcing Strategy and Sustainable Competitive Advantage", Industrial Marketing Management, Vol. 33, No. 1, pp. 7–14.

Lall, S. (2000) Strengthening SMEs for international competitiveness, ECES Working Paper 44, Egyptian Center for Economic Studies, Cairo.

Lee, W.B. and Lau, H.C.W. (1999) "Factory on demand: the shaping of an agile production network", International Journal of Agile Management Systems, Vol. 1, No. 2, pp. 83–87.

Lester, R.K. (2003) China, America, and the Global Competition for Industry, MIT IPC Working Paper IPC-03-005, Massachusetts Institute of Technology, Cambridge, MA.

Maropoulos, P.G. (2003) "Digital enterprise technology - defining perspectives and research priorities", International Journal of Computer Integrated Manufacturing, Vol. 16, No. 7–8, pp. 467–478.

McGuire, J. and Dow, S. (2003) "The persistence and implications of Japanese keiretsu organization", Journal of International Business Studies, Vol. 34, No. 4, pp. 374–388.

Medcof, J.W. (2001) "Resource-based Strategy and Managerial Power in Networks of Internationally Dispersed Technology Units", Strategic Management Journal, Vol. 22, No. 11, pp. 999–1012.

Micklethwait, J. and Wooldridge, A. (1996) The Witch Doctors - What the Management Gurus are Saying, Why it Matters and How to Make Sense of it, Heinemann, London.

Milberg, J. and Dürrschmidt, S. (2002) "Planning Methodology for Changeable Logistic Systems", Production Research, Vol. 9, No. 1, pp. 63–68.

Miles, R.E. and Snow, C.C. (1984) "Fit, Failure and the Hall of Fame", California Management Review, Vol. 26, No. 3, pp. 11–28.

Mody, A. (1993) "Learning through alliances", Journal of Economic Behaviour and Organization, Vol. 20, No. 2, pp. 151–170.

Molina, A., Mejía, R. and Velandia, M. (2003) "Core processes, methods and e-services to support Virtual Enterprise Brokerage", International Journal of Networking and Virtual Organizations, Vol. 2, No. 1, pp. 33–49.

Mraz, R. (1997) "Building a Manufacturing Network", Machine Design, Vol. 69, No. 16, pp. 85–89.

Nagel, R.N., Dove, R., Goldman, S. and Preiss, K. (1991) 21st Century Manufacturing Enterprise Strategy: An Industry Led View, Iacocca Institute/Lehigh University, Bethlehem, PA.

Noori, H. and Lee, W.B. (2002) "Factory-on-demand and smart supply chains: the next challenge", International Journal of Manufacturing Technology and Management, Vol. 4, No. 5, pp. 372–383.

O'Neill, H. and Sackett, P. (1994) "The Extended Manufacturing Enterprise Paradigm", Management Decision, Vol. 32, No. 8, pp. 42–49.

Penrose, E.T. (1963) The Theory of the Growth of the Firm, Basil Blackwell, Oxford.

Porter, M.E. (1990) The competitive advantage of nations, Free Press, New York.

Prahalad, C.K. and Hamel, G. (1990) "The Core Competence of the Corporation", Harvard Business Review, Vol. 168, No. 3, pp. 79–91.

Probert, D. (1997) Developing a make or buy strategy for manufacturing business, Institution of Electrical Engineers, London.

Riedel, J. and Pawar, K.S. (1998) "The strategic choice of simultaneous versus sequential engineering for the introduction of new products", International Journal of Technology Management, Vol. 6, No. 3/4, pp. 321–334.

Ritter, T. and Gemünden, H.G. (2003) "Network competence: Its impact on innovation success and its antecedents", Journal of Business Research, Vol. 56, No. 9, pp. 745–755.

Robertson, P.L. and Langlois, R.N. (1995) "Innovation, Networks, and Virtual Integration", Research Policy, Vol. 24, No. 4, pp. 543–562.

Roosendaal, C. (2000) The Concept of the Virtual Organisation, Delft University of Technology, Delft.

Ross, E.M. (1994) "The Twenty-first Century Enterprise, Agile Manufacturing and Something Called CALS", World Class Design to Manufacture, Vol. 1, No. 3, pp. 5–10.

Sadd, R.J. and Bennett, D.J. (1999) "Automotive Supply Logistics: Myth and Reality", in Proceedings of the 4th International Symposium on Logistics, Servizi Grafici Editoriali, Padova, Italy, 11–14 July.

Shi, Y. (2003) "Internationalisation and evolution of manufacturing systems: classic process models, new industrial issues, and academic challenges", Integrated Manufacturing Systems, Vol. 14, No. 4, pp. 357–368.

Shi, Y. and Gregory, M. (1998) "International Manufacturing Networks - To Develop Global Competitive Capabilities", Journal of Management, Vol. 16, No. 2/3, pp. 195–214.

Skinner, W. (1969) "Manufacturing - missing link in corporate strategy", Harvard Business Review, Vol. 47, No. 3, pp. 136–145.

Skinner, W. (1974) "The Focused Factory", Harvard Business Review, Vol. 52, No. 3, pp. 113–121.

Skinner, W. (1986) "The Productivity Paradox", Harvard Business Review, Vol. 64, No. 4, pp. 55–59.

Smulder, F.E., Boer, H., Hansen, P.H.K., Gubi, E. and Dorst, K. (2002) "Configurations of NPD - Production Interfaces and interface Integration Mechanisms", Creativity and Innovation Management, Vol. 11, No. 1, pp. 62–73.

Souder, W.E. and Nassar, S. (1990) "Choosing an R&D Consortium", Research Technology Management, Vol. 33, No. 2, pp. 35–41.

Spring, M. and Dalrymple, J.F. (2000) "Product customisation and manufacturing strategy", International Journal of Operations & Production Management, Vol. 20, No. 4, pp. 441–467.

Stein, T. (2002) "Keiretsu Comedown", Investment Dealers' Digest, Vol. 68, No. 6, pp. 28–36.

Stock, G.N., Greis, N.P. and Kasarda, J.D. (1999) "Logistics, strategy and structure: A conceptual framework", International Journal of Physical Distribution & Logistics, Vol. 29, No. 4, pp. 224–239.

Sturgeon, T.J. (2002) Modular Production Networks: A New American Model for Industrial Organization, Industrial Performance Center, Massachusetts Institute of Technology, Cambridge, MA.

Tesar, G., Boter, H. and Bohman, H. (2003) "Network participation and internationalisation of smaller manufacturing enterprises in Central Europe: A Case for Research", in Proceedings of the 19th Annual IMP Conference, Lugano, 4–6 September.

Tian, G.Y., Yin, G. and Taylor, D. (2002) "Internet-based manufacturing: A review and a new infrastructure for distributed intelligent manufacturing", Journal of Intelligent Manufacturing, Vol. 13, No. 5, pp. 323–338.

Tidd, J. (1995) "Development of Novel Products Through Intraorganizational and Interorganizational Networks", Journal of Product Innovation Management, Vol. 12, No. 4, pp. 307–322.

Uzzi, B. (1997) "Social Structure and Competition in Interfirm Networks: The Paradox of Embeddedness", Administrative Science Quarterly, Vol. 42, No. 1, pp. 35–67.

Vaan, M.J.M. de, Verweij, M.J., Laan, J.van der and Wigman, J. (2002) Voorstudie Innovatief Onderzoeksprogramma 'Integratie in Ontwikkeling and Productie in de Maakindustrie', Berenschot, Utrecht.

Wiendahl, H.-P. and Lutz, S. (2002) "Production in Networks", Annals of the CIRP, Vol. 51, No. 2, pp. 1–14.

Zachary, K. (2001) "Dismantling the Keiretsu", China-Britain Trade Review, Vol. 37, No. 5, pp. 14–16.

PART I:

Networks as Complex Adaptive Systems

Prospects for Industrial Networks

Chapter 2 has proclaimed that views on industrial networks have changed and that this brings about challenges in comprehending the interactions. Particularly enabled by ICT and increasing specialisation, networks have become common ground in industry, whether it concerns Strategic Networks (Section 1.1) or networks of loosely connected entities (e.g. Dispersed Manufacturing Networks). Strengthened by the four shifts, identified by Bennett and Dekkers:

- manufacturing capacity becoming a commodity
- increasing specialisation of individual firms,
- resource allocation linked to capital investment and manufacturing as commodity, and
- more loosely connected agents,

networks are moving away from traditional concepts. These traditional concepts, like keiretu and chaibol networks that can be considered as an extension of Strategic Networks, do not offer the responsiveness and reach of more loosely connected networks, e.g. as found in Dispersed Manufacturing Networks. This calls for appropriate approaches to reconfiguration and resource allocation, and for understanding of the principles behind horizontal and vertical collaboration in the loosely connected networks. Especially in collaboration, one issue dominates: networks are about interactions between agents.

Introducing Part I: Networks as Complex Adaptive Systems

That interaction constitutes the realm of complex systems theory applied to industrial networks; hence, the next three contributions offer an expansion of this perspective of complex systems, albeit from different starting points. The first contribution in this part of the book by Hamid Noori and W.B. Lee offers two perspectives on industrial networks. Firstly, they view Dispersed Manufacturing Networks as an archetype on how smaller firms may collaborate and thus compete at a global scale. This perspective is not new, e.g. Chetty and Campbell-Hunt (2003, p. 800) point to the network strategy as one possible route for internationalisation, but the chapter expands on solutions for managing these networks. The second perspective considers Dispersed Manufacturing Networks as Complex Adaptive Systems, similarly to Biggiero (1999) and Andriani (2003) for regional networks in Italy. Hermann Kühnle in Chapter 4 builds on the proposal for the behaviour of complex systems by adding self-criticality and similarity as essential ingredients; e.g. Song *et al.* (2005) consider self-similarity as a keystone for scale-free networks. In Chapter 5 Rob Dekkers offers an outlook on how to combine this complex systems view with evolutionary models, co-evolution, game-theoretical approaches and network theories. During the years to come, we might expect that further elaboration of the complex systems view in its widest sense will add to the understanding of agents' behaviour in industrial networks (e.g. Iansiti and Levien [2002, pp. 55-58] and Surana *et al.* [2005] follow similar reasoning) and to the improvement of coordination mechanisms between loosely connected entities. The three contributions address both practical issues and underpinning theory.

References

Andriani, P. (2003) "Evolutionary Dynamics of Industrial Clusters", in: E. Mitleton-Kelly (Ed.) Complex Systems and Evolutionary Perspectives on Organisations, Pergamon, Oxford, pp. 127–148.

Biggiero, L. (1999) "Market, hierarchies, networks, districts: A cybernetic approach", Human Systems Management, Vol. 18, No. 2, pp. 71–86.

Chetty, S. and Campbell-Hunt (2003) "Paths to internationalisation among small- to medium-sized firms: A global versus regional approach", European Journal of Marketing, Vol. 37, No. 5/6, pp. 796–820.

Iansiti, M. and Levien, R. (2002). The New Operational Dynamics of Business Ecosystems: Implications for Policy, Operations and Technology Strategy, Working Paper 03-30, Harvard Business School, Boston.

Song, C., Havlin and Makse, H. (2005) "Self-similarity of complex networks", Nature, Vol. 433, No. 7024, pp. 392–395.

Surana, A., Kumara, S., Greaves, M. and Raghavan, U.N. (2005) "Supply-chain networks: a complex adaptive systems perspective", International Journal of Production Research, Vol. 43, No. 20, pp. 4235–4265.

3 Dispersed Network Manufacturing: An Emerging Form of Collaboration Networks

Hamid Noori
Wilfrid Laurier University

W.B. Lee
The Hong Kong Polytechnic University

Abstract In this chapter we introduce a conceptual framework for a new form of production system, which is unique from many perspectives. The proposed framework, which we refer to as Dispersed Network Manufacturing (DNM), is based on the creation of a network of plants that are electronically linked so that the participating members focus on their specialized tasks, yet also share their manufacturing and production resources to create a loosely structured and flexible enterprise.

Keywords Complex Adaptive System, Dispersed Manufacturing Networks, Manufacturing paradigm, SMEs

3.1 Introduction

The new global economy presents many opportunities and dilemmas to many companies around the world. While many theories (e.g. Organizational Ecology by Hannan and Freeman [1989]) have been stated to help understanding the global environment in which Small- and Medium-sized Enterprises (SMEs) compete, we feel that the essence of those concepts mostly expects SMEs to act in a vassal role for big companies, painstakingly making profit by being a subcontracting member of a larger manufacturing network. This is so because traditionally SMEs either select market segments that require no head-on competition with the giant companies' technological strength, or live with the big companies by working as the attentive members in their supply chain. These phenomena will only be intensified as more market-entry freedom is granted to those international big corporations where local government protections and subsidies are prohibited. But, are there other ways for SMEs to come together to compete in the globalized arena?

It is well understood that the structure of alliance networks influences their potential for knowledge creation (Schilling and Phelps, 2007). In a global context, how can SMEs, spread geographically in a certain region, form a dynamic and adaptive network to create competitive advantages on both collaborative and individual scales? In this chapter we discuss a completely different form of collaboration between SMEs than what has already been discussed in the literature. Our position is not to re-assert any traditional collaborative network concepts nor is it to examine an economic world that is dominated by critical masses in one place. Rather, we propose that SMEs may steal the flash from the big companies in the global village when appropriate technological and infrastructural developments are pursued.

In support of this proposition, we believe that the prototype of a radically different solution can already be found in the Pearl River Delta (PRD) region of China (Enright *et al.*, 2005). The industrialization of the PRD has been principally led by Hong Kong manufacturers and primarily by its SMEs, who throughout the years have developed the skills to deal with multiple cultures and a globalized business environment. In doing so, Hong Kong SMEs have been able to develop a reciprocal bonding, which calls for no obligatory egalitarian responsibility to each other – a notion that we refer to as Dispersed Network Manufacturing (DNM).

The objective of this chapter is to provide a better understanding of the notion of DNM as a new business model and to demonstrate the Dispersed Manufacturing Network (DMN) as its possible realization. DMN provides a prototype as how SMEs can form a dynamic and adaptive network to create competitive advantages on both collaborative and individual scales. The proposed system is based on the creation of a network of plants that are electronically linked so that the participating members focus on their specialized tasks yet also share their manufacturing and production resources to create a loosely structured and flexible enterprise. We start our discussion with a look at how SMEs typically connect which, in our view, is a preamble for understanding effective operations on the global stage.

3.2 The Role of SMEs in the Economics of Competition

In the present literature, a considerable number of studies have focused on two separate but interrelated issues: (i) the impact of globalization on manufacturing operations and the roles SMEs could play therein, and (ii) the technical and organizational challenges of integrating activities along the upstream–downstream value chains (see Noori and Lee [2002] for more details). In practice, the extent of integration must meet a number of requirements including addressing multiplant operations, handling Supply Chain Management and distribution, integrating financial analysis, providing real-time forecasting, supporting Manufacturing Execution Systems and finite scheduling, and providing executive information and decision making tools.

This change of perspective from the lean factory to a collection of lean and agile world-class decentralized facilities able to work together as an "extended enterprise" (O'Neil and Sackett, 1994) using common information technologies is now emerging and is being discussed by various researchers. For example, Hanson and Voss (1999) state that the vision of the future manufacturing enterprise configuration is based on two principles: (a) the portfolio of manufacturing facilities that may not all be directly owned, but which add up to a complete capability, and (b) the presence of a global brokerage, which has direct and intimate customer relationships and can conceive total business solutions that are tailored to customer needs.

In this knowledge-based environment, the driving forces are digitization, the Internet, and high-speed data networks that are keys to addressing many of the operational issues from design to logistics and distribution.

Given the current understanding of the multinational corporation (Noori, 1998), it is an interesting question to pose: What would happen if the world of a global company were reversed so that instead of power resting solely with singular entities there was the sharing of it among a collection of small players? How would the rules change?

The era of the globalization of manufacturing and advances in microelectronics has brought about further challenges to the implementation of the enterprise integration concept (Noori and Mavaddat, 1998). In this context, the roles of SMEs have been discussed from different perspectives, like cultural as in "Keiretsu" in Japan (Noori and Radford, 1995) or "Chaebol" in South Korea (Dyer *et al.,* 1998), or like community entrepreneurship as in the Italian leather fashion cluster (Porter, 1998), and like "dynastic enterprise" and "borderless manufacturing" as in the case of Li & Fung in Hong Kong (Magretta, 1998).

In particular, it has been argued that the emergence of loosely tied and fragmented production systems (Moran and Riesenberger, 1994) and the new pattern of inter-firm relationships are evolving into a different form of enterprise-wide integration and co-operation beyond what has been seen so far. The objective is the creation of a network of plants that are perhaps geographically dispersed but electronically linked so that the participating members focus on their specialized tasks yet also share their manufacturing and production resources to create a loosely structured and flexible enterprise.

Foremost, we need to clarify the meaning of "Dispersed Manufacturing" or "Distributed Manufacturing" (Montreuil *et al.*, 2000) that has been used in the literature with a variety of meanings and contexts. At one extreme of the "spectrum" of meaning (which certainly has more than one dimension) we have dispersion, essentially referring to the supply chain being spread over a worldwide network. A great example of this form of supply chain (or, more accurately, supply net) management is that of Li & Fung (Magretta, 1998). In that case, due to the sheer size and power of the corporation, a network of thousands of globally placed SMEs exists, which can be quickly organized to meet particular production planning, organizing, and management needs. While flexible and highly agile, the network exists because of the reputation of the corporation and the trust it has with its suppliers in making high-volume orders. The meaning of "dispersion" in this case is synonymous with "lacking geographical borders".

Another extreme of the spectrum is the type of distribution arising from geographically localized community-based initiatives, in which a collection of smaller players organizes itself (i.e. forms clusters) to provide to customers a high-value and highly customized supply chain function that would have been unattainable by any of the members alone (Porter, 1998). One modern example of inter-firm, intra-regional industrial networking comes from the US state of Iowa. The idea was the result of several machining and foundry shops recognizing that if they pooled their capabilities into a regional area manufacturing partnership, they would be able to collectively bid on OEM outsourcing contracts that would have otherwise been out of reach. Thus experienced and financially sound SMEs could provide their services to a large number of OEMs. The Iowa Model has since expanded into a limited liability hub that generates leads, coordinates and schedules orders and product flows, and provides services and overall management. Furthermore, as the network continues to grow the number and size of bids has expanded, presenting a new scope of market opportunities. Additionally, there are a number of smaller firms marketing proprietary products, and suppliers within the alliance carry third-party registration to such standards as ISO, AS9000, etc. This model provides a good example of the power of a collaborative network as an organizational agent aiding production.

Dispersed Network Manufacturing in the context of this chapter explores a different paradigm from the previous two cases. It attempts to draw the flexibility of the large-scale global network and the self-organization of the multi-member community into a highly customized self-adapting network structure that moulds itself to the resources and constraints of the local business environment. Such networks are highly flexible in responding to current market needs. This strategy allows SMEs to leverage their combined capabilities in a dynamic way that strengthens each individual unit through a dynamic competitive process that selects for the best possible end-product and most efficient production path for the customer (for further discussion, see the appendix to this chapter).

In the remainder of this chapter, our focus will be on DNM as an emerging business model, and DMN as prototype examples of such a business model. In this context we will also explore the notion of hubs and the role they play in forming Dispersed Manufacturing Networks. See Figure 3.1 for a schematic view of a Dispersed Manufacturing Network.

3.3 Characteristics of Dispersed Network Manufacturing

As mentioned, DNM is a manufacturing philosophy that allows a set of geographically separated (but likely regional) SME partners to come together and form a complete production capacity that can be sold to a larger customer. A DMN, on the other hand, is the name given to a particular combination of SMEs that has formed in accordance with the principles of DNM. It is a dispersed network topography that has been modelled to suit particular manufacturing purposes.

DNM inherits its ideological roots from "community entrepreneurship", described above, but at the same time allows for larger scaling, thus affording competition with conglomerate firms. SMEs form dynamic networks to meet a specific order request, and then dissolve as necessary. Customers place orders onto the network, which subsequently form community clusters that compete to win the order bid (or portions of the contract).

The key for SMEs is to be able and willing to adapt themselves continuously as the business environment changes from customer to customer. The notion of adaptation here is far more drastic than the usual notions of flexibility, agility, and rapid response. It has been noted before that adaptation is the appropriate response

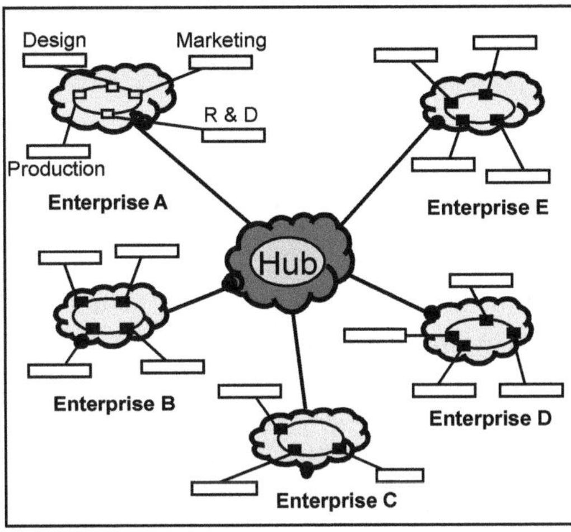

Figure 3.1. *A schematic view of a typical Distributed Manufacturing Network*

to change (Haeckel, 1999). To this end, Haeckel has previously described the terms adaptive enterprise, and sense-and-respond in describing how organizational behaviour should respond to change. In our opinion, DMNs resemble "sense-and-respond" organizations, but require a more literal reorganization of the production chain with each new customer order. The virtual firm, networks and cross-functional teams are all usually associated with adaptive DMNs. Some of the characteristics associated with DMNs resemble those of agile organizations (Shaw and Lengyel, 1996) and are listed Figure 3.2.

Most of the terms employed in the figure are recognizable and their use is certainly not limited to DNM applications. The lists of attributes and enablers

DMN Enterprise Attributes

- Adaptive Organization
- Collaborative Production
- Knowledge-driven Enterprise
- Solutions Provider
- Niche Marketer: Many, Changing Models
- Individualizable Products and Services
- Arbitrary Production Volume
- Production to Order
- Enduring, Interactive Customer Relationships
- Proactive Marketplace Change Agent
- Others

DMN Attribute Enablers

- Enterprise-wide Integration of Processes/ Systems
- Intercompany Integration of Processes/Systems
- Neutral Databases for Information Storage and Retrieval
- Enterprise-wide Performance Assessment Models/Metrics
- Information Exchange Standards
- Digital Product Description Standards
- Reusable Design Tools
- Reusable Production/Business Processes
- Flexible, Modular, Reconfigurable Production Technologies
- Cross-functional, Entrepreneurial Teams
- Physically Distributed Teams, Intra- and Intercompany
- Robust Groupware
- Interoperable Simulation and Modeling Processes/Tools
- Real-time Production Management/Scheduling
- Agility-based Quality Metrics
- Real-time Information Access
- Others

Figure 3.2. *DMN attributes and enablers (adapted from Shaw and Lengyel [1996])*

describe only some of the aspects of DMN, yet provide a guideline for what it means for companies to become part of such a system. They could also be used as a method to measure or audit the individual DMNs.

3.4 Dispersed Network Manufacturing as a Complex Adaptive System

A Complex Adaptive System (CAS) is the study of natural systems – how they adapt, interact, and survive over time. Just as biological systems are governed by Darwinian natural selection, which favours those agents that successfully adapt to the (dynamic) constraints and resources offered by their physical environment, so too will any "organic" theory of business network formation be subject to a mechanism that selects for those of its members that co-evolve, and that successfully adapt to the environment (as set by globalization) and the available resources (various manufacturing and strategy inputs, for example). Complex Adaptive Systems Theory attempts to categorize the common characteristics of complex systems wherein individual agents interact and compete for resources within their environment. These agents can aggregate to form more complex multi-agents which often contain highly specialized component agents that might individually have little in common with the function of the overall system. In Complex Adaptive Systems small changes are not ignored and that "emergence" not "planning" is how things get done.

The treatment of DNM as a Complex Adaptive System has strong parallels to literature on industrial districts and networks (e.g. Biggiero, 1999; McKelvey, 1999; Newman, 2003; Pietarinen, 2004; Robertson and Langlois, 1995). Some key points of CAS are that they contain "lever points", critical nonlinear features where a small input signal can be magnified to impact the entire system (such as a small quantity of pathogen leading to a macro-scale response within the immune system, the principle that underlies vaccines). Furthermore, CAS agents exhibit the capability of foresight: they interact, combine, and replicate by means of an internal (endogenous) mechanism for determining the fitness of the various agents and their interactions (see Dekkers, 2005; Holland, 1995). Those agents (or agent aggregates) that are more fit recombine (reproduce) more successfully than those that are less fit, following a sort of natural selection. The fundamental objective in defining CAS is to describe the properties of the agents and their behavioural abilities; the selection process must follow naturally, a sort of "autopilot" that does not rely on direct external (exogenous) control, though such control can certainly be used to impact the environment in which the agents exist.

Strategy in the network is directly related to the strategies that agents use to compete for scarce resources. If necessity is the mother of invention, then scarcity is the mother of innovation and adaptation. Given that we intend to discuss the structure of self-organizing clusters of SMEs capable of forming a cohesive manufacturing strategy, drawing an analogy with how complex natural networks adapt to their environment using the available resources would shed light on how

DNM might operate (see Kauffman [1994] for a related discussion and links to further resources).

As will be elaborated in the following section, competition is different for DMNs than for more traditional supply chain structures. The goal of DMNs is self-organization and dynamism in the form of continual adaptation to changing circumstances. When a customer order specification is advertized to the SMEs on the network, particular production units combine to create a competitive production capacity for a portion (or all) of the job order. In this milieu, management becomes not production-oriented but mission-oriented. At the same time, hierarchy dissipates and dissolves, since to be effective the participative companies need free rein to remain flexible and responsive. The DNM business system demands flat hierarchies, mission orientation, and above all a sense of direction.

3.5 Designing DNM as a Business System

One of the main characteristics of DNM is its ability to manufacture a high variety products (i.e. mass customization) in a production environment where the emphasis is on both flexibility and speed. This is consistent with Pine's (1992) argument that mass customization is best suited for a dynamic and uncertain environment. More uncertainty implies the need for a higher degree of customization. On the other hand, as the number of product options increases, and the number of supply chain choices expands, achieving mass customization becomes a daunting task due to the challenges of matching product supply with consumer demand.

What if, instead of making a vain attempt to control this chaotic supply environment in a top-down fashion, the supply chain were asked to "sort itself out," to understand its own logistics problems on the micro-scale and find solutions that would bubble up to the macro-world. There is much in the literature to suggest that dynamic, adaptive systems work best at the "edge of chaos", the critical density point on the cusp of order where there is sufficient uncertainty to allow for the exploitation of those remote opportunities and uncertain connections that are probabilistically minimized in an ordered environment (Kauffman, 1994). Though these ideas have a formal rooting in mathematics, they convey certain qualitative heuristics that can be applied to the DNM business model.

One such heuristic rule is that DMNs work best in the presence of noise and uncertainty, not meaning that management is impossible but that a well-defined a priori top-level organizational structure is overly rigid and thus unnecessary (or perhaps even counter-productive) to the efficient operation of the network. Thus imagining a network, some central "pseudo-controller" known as a hub is necessary to introduce SMEs together on a large enough scale to foster meaningful communication warranting the cost of maintaining the hub. For an apt analogy, consider the World Wide Web, with its network of online users. Though each user is technically able to communicate with any other, this is rarely done directly unless the users have been previously introduced. Thus when disparate users wish to connect to exchange information, they go to a hub represented by a search engine portal such as Google, Yahoo, and MSN, or they go to a community chat room or interactive blog space.

So, in the context of DNM, the role of a hub as a pseudo-central controller is to provide a common technological forum to introduce various SMEs and facilitate information exchange, but its role in explicitly coordinating production would be rather limited. It actually does very little controlling, which is done co-operatively by the SMEs themselves, and serves more as a co-ordinating conduit. In other terms, the hub takes on the role of a global brokerage (Hanson and Voss, 1999) with regards to interconnecting clients and SMEs. Potential customers communicate with the SMEs via the hub. So while in theory any SME could interact with any other, in practice each would only cluster with several partners (somewhat similar to Porter's sense) to form a complete production capacity, and it is not necessary for a knowledge of the entire production chain outside of the cluster to be understood. Only the ultimate order-placing customer who chooses among the available clusters for production will have this knowledge. It is not that the complete production chain need be a secret; it is that the individual clusters have no need for this information. They must only know how to co-ordinate to deliver on their stated production capacity.

Another heuristic rule applies directly to the SME as an individual business. Each SME must be free to change its internal strategy and goals as it adapts to and grows with the changing business environment. The lesson to be drawn from Darwinian natural selection is that, ultimately, it is the adaptability and success of the individual that is paramount, not that of the group per se. Thus "Survival of the Fittest" must apply to individual SMEs, and not to the Porter-esque clusters to which they may belong. Clusters become more effective over every successive generation of the business model precisely because the strong SMEs survive and are made stronger while their weak counterparts are weeded out and perish. Therefore it cannot be stated that DNM is responsible for turning a member SME into a winner: the onus of success forever remains with the individual firm, whereas the business model can only provide all the opportunities to take advantage of the current environment.

These two characteristics (i.e. decentralized control and dynamic internal adaptation) in particular are keys to obtaining true modularity in the business model. Decentralized hub control (known in technical terms as scale-free networking) combined with a selfish improvement standard ensure that, no matter how the DMNs come and go with future customer orders, each SME partaking in the network will be as fit as possible. Those that cannot adapt to changes and maintain an ever evolving standard of fitness will be weeded out. This implies that those SMEs that participate in the DNM structure will be more responsive to market demands (vis-à-vis those that do not), as they will be in a position to form clusters that meet the precise needs of the moment. Traditionally, an SME may make itself more efficient in response to a current production strategy, but when another strategy is required it may find itself hard-pressed to adapt its methodologies to the new constraints. DNM allows SMEs to select, preserve, and strengthen those business processes that are considered positive, while each new cluster formation (i.e. each new generation) gives the opportunity to slough-off those deemed inefficient or counter-productive, naturally subject to the particular constraints.

Organizationally, a DMN is a network (web) of independent partners (SMEs) with their own core competencies whose presence in the web is based on an economically valid value proposition. Through time, such a web is not likely to remain stable due to changing technology, customer needs, and societal conditions. Therefore, DNM depends on a dynamic, networked approach to managing and communicating information. This is an organization that promotes both efficiency of production and speed in responding to customer needs. It is a hybrid of traditional hierarchical organization and traditional entrepreneurial organization that addresses both complexity and speed simultaneously (see Applegate [1999] for a related discussion). Its benefit to SMEs is primarily in expanding their economic outreach in more and bigger contracts, enhancing opportunities for knowledge diffusion and new capabilities formation through close and varied partner relationships, and providing stability by spreading business partnerships over a wide range of producer locations and competencies. In such a fashion, new opportunities are better assessed and exploited by individual SMEs through the tools they have obtained via collaboration.

3.6 Transitioning from Traditional Network Structure to Dispersed Network Manufacturing

An important question yet to be addressed is how to make the transition from a more traditional supply-chain network structure to the more distributed nature called for by DNM. One idea is to look for inspiration from the structure of the World Wide Web itself, to consider the notion of scale-free networks (Li *et al.*, 2004). Such networks are characterized by many members, or nodes, any of which is theoretically capable of communicating with any of the others. However a few principal nodes, called "hubs", are connected to exponentially more nodes than the vast majority of the others, which have a relatively low degree of connectivity – see Figure 3.3. Scale-

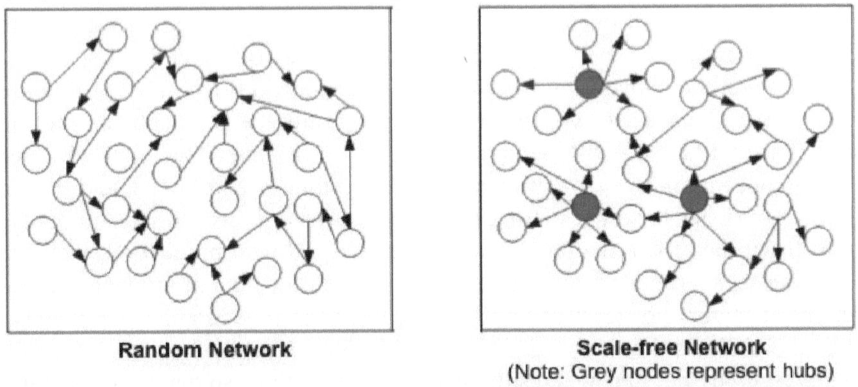

Figure 3.3. *Graphical comparison of a "Scale-free Network" with a "Random Network" and the role of hubs (adapted from: http://en.wikipedia.org/wiki/Image:Scale-free_network_sample.png)*

free networks are very resilient to random malfunction, in the sense that the majority of their nodes can become damaged yet the network function persists (see Li *et al.*, 2004). On the other hand, they are very easily compromised by hub failure; that is should the hubs fail in simultaneous fashion, and then because of their high degree of connectivity the network itself will fail. What this means for the DNM model is that there are no critical SMEs in the supply net; even if a large number go bankrupt the model would still be viable. The hubs are either extremely well connected and powerful SMEs (a doubtful proposition as this would simply become another version of the dynastic enterprise characterized by Li & Fung), or else technology portals used to facilitate information exchange between customers and SME. These would be maintained at mutual cost to all parties involved by an associated partner. Hubs are critical to the DNM model and cannot be shut down other than at great expense to the entire system.

The point to all this, then, is to recognize hubs as the centralized managers capable of bringing the disparate production partners together for a specific job. Without them, SMEs would simply operate according to the classical supply chain models. The fact that hubs easily bring many nodes together makes them a natural point of "control" for the system, thus making them the "go-to" nodes for those wishing to make a connection to the production network . Either the hubs are mutually owned and maintained by all partners or they are founded by an external entity such as an academic operator or a private firm. In the latter case, there must be more than one hub so that no one "tyrannizes" the others. There must also be more than two hubs so that they do not slit each others' throats in competition. The purpose of the hubs is not to be competitively involved in the production process but to facilitate competition among the SMEs in bidding for customer contracts. There must be many hubs with many redundant connections so that there is more than one way to link members to execute a given task; this is the only way to safeguard the network against hub failure.

Implicit in the design of such a production network is a unified system of communication whereby simultaneous order information can be routed through the hubs to the multitude of branch SMEs. This is a technology issue. Should such a system be realized, then there would be a consistent way of disseminating an order request through the network by passing it to the connection-rich hubs. A modified version of the digital Factory-on-Demand (FOD) prototype is potentially one such system (Lee and Lau, 1999; Noori and Lee, 2002). Ultimately, the hubs will focus the main concentrations of wealth between the customers and SMEs, and the rich will only get richer for orchestrating the business transactions – it pays to manage a hub! It will be re-iterated that hubs are not involved with production; they only co-ordinate asynchronous information transactions. They gain prominence through a variety of factors not least of which being the first-mover status. Ultimately it will be those SMEs that are most fit in their dealings with the hubs that can draw enough resources and thus make sufficient new connections so that they will thrive as well. As each new incoming project offers the chance to reaffirm those processes that were proven highly effective, while discarding those that are deemed ill-suited, the fittest SMEs are able to sustain themselves through this process of continual renewal and adaptation to the most pressing market requirements. Given that any business unit will have deficiencies in a certain area, collaboration can be used as a method to draw resources into that capabilities gap. Not only does this address weaknesses in the production process, but it also accounts for a form of knowledge diffusion that enhances the performance of an SME as a unique enterprise, thereby bolstering its ability to operate within a DNM environment. Thus collaboration in such a network formation prepares small producers to better exploit opportunities in the market while sustaining their growth by drawing in the various resources so critical to new capabilities formation.

So the real question thus becomes, who co-ordinates the formation of a particular DMN for a given job? The answer is paradoxical: no one and every one. There is no central command; the job requirements are distributed to all nodes via the hubs using a common technology language and interface (e.g. a FOD). The nodes will then be able to determine on their own whether they have enough resources and connections (i.e. can some nodes come together on their own to form a cluster) to manufacture a particular component of the design specification or to meet a certain design requirement or facilitate a certain design transaction. Then this capabilities information is passed back to the hub again using the common technology interface where the customer can then assess the available options and select among viable clusters.

It is important to note that the hub only facilitates communication and information sharing between customer and SME, and between SMEs themselves. The hub itself does not determine the formation of clusters (this is done by the SMEs themselves) or the awarding of job contract to a cluster (this is done by the customer). The benefit of this method is that all information from the available clusters is made available to the customer; should a chosen cluster underperform or be otherwise unable to meet the stated requirements, another cluster can be chosen as necessary. This dynamic nature of the system means that it is asynchronously regulated, while information is circulated internally through ad hoc feedback and feedforward loops. Clusters can continuously form/reshape themselves and re-advertize their abilities to the hub; they are highly customizable to the constraints of the business environment and hence contribute immensely to overall enterprise agility. In fact certain dominant or niche clusters could be continuously posted to the hub even if no specific job request demands their existence; but this is how new or strong capabilities could be advertized. Thus SMEs are not dependent on large corporations for control and leadership of their production processes, but rather only for order placement. The big companies are needed as order placers such that SMEs could build-up confidence in a DNM environment by working together on a large project. This is also how the effective network relations are formed and made persistent. After certain dominant or niche clusters have formed over several generations of DMN, they may find themselves able to challenge larger producers with the combined, effective, and highly streamlined production capacity. It is this capacity that can be now advertized to potential customers via the hubs.

Subsequently each cluster would operate on its own to do its job share, measuring performance and trying to maximize its own fitness. Performance fitness is defined endogenously (i.e. not externally) to the system; it is not an explicit metric stated academically, but a (quantifiable) performance measure that ultimately means, simply, how a given SME can procure more jobs, get more work with this cluster or maybe even with another cluster.

The conclusion, therefore, is to adopt a somewhat laissez faire approach to this issue of network control. Assuming that there are enough SMEs connected to a hub, customer order specifications need only be broken down and advertized on the hub, and the SMEs themselves will form clusters and bid competitively on the available work.

3.7 A Typical Dispersed Manufacturing Network at Work

Let us look at a prototype example of how DNM might function in reality. We begin with a model universe, which consists of many SMEs and a few hubs, all connected together with a common FOD-type communication/interaction technology. Factory-on-Demand is a technological system for co-ordination information sharing, supply and production chain functions, cluster formation, and enterprise co-ordination (see Figures 3.4 and 3.5 for schematic representations of enterprise and technology relations, respectively). It draws heavily from the object-oriented design philosophy of computer science and the multi-agent logic of Complex Adaptive Systems (CAS) theory. The system facilitates the benchmarking of product information as well as order placement and supply co-ordination. The largest achievement of FOD is in its ability to dissect production workflow based on real-time customer input, simultaneously considering all available resources. The customer is able to convey product specifications to the system, and the requirements are subsequently broken down such that an analysis of the required production capability, process plan, delivery schedule, and other value chain information may be constructed. This allows for the production of small quantities with dissimilar design, a key consideration

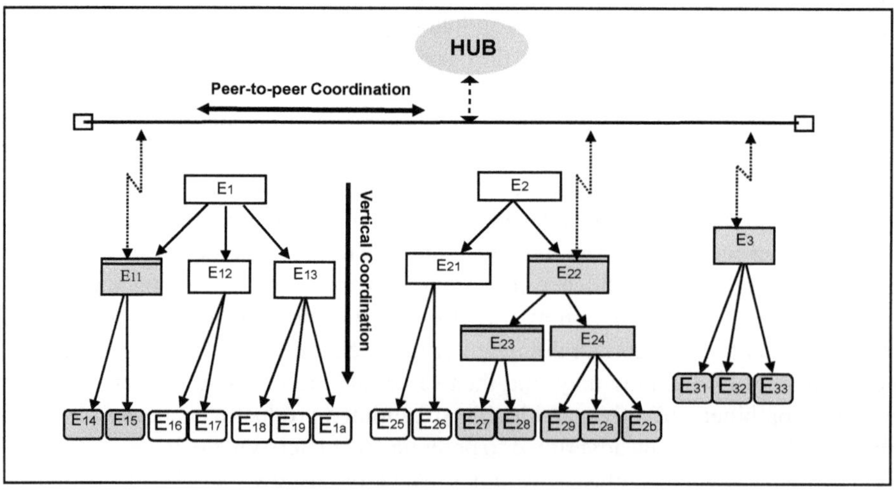

Figure 3.4. *Factory-on-Demand enterprise relations (adapted from Lee and Lau [1999])*

in mass customization. SME clusters are subsequently able to synthesize this information and provide feedback to the customer regarding production capabilities and order confirmation. The ultimate goal of the FOD system is to provide the right information to the right place at the right time and in the right format throughout the inter-enterprise concurrent engineering process.

Back in our model universe, say a large enterprise named MegaCorp wants to place a job order with a capable group of contractors for various segments of its product design; it does not know in advance who it will select for the job, it may not even have an idea about potential manufacturers. What MegaCorp can do is sign onto a DNM hub and advertize its order there, subsequently disseminating the information among the relevant SMEs. Now the SMEs have their manufacturing capabilities openly advertized via the FOD technology through the hubs, so each has an understanding of what the others is capable of doing (or at least what they claim to be capable of). When the order requirements are known to the SMEs, based on whatever subjective decision-making strategy (such as geographical proximity, capabilities complementarity, reputation, etc.), small "clusters" (or "cells", in network theory jargon) will join together to form a complete capability function which is then published on the hub. Other SMEs will form clusters of their own, and these clusters will then compete in a bidding process to win the job from MegaCorp. Note that each of these clusters is not entirely similar to the clusters suggested by Porter (1998) as they are somewhat more flexible. In this case, each cluster need not be a permanent entity (most certainly not the bid losers, anyway), but one that may dissolve and reform as the jobs come and go. Ultimately the winning cluster gets the job and delivers the goods – see Figure 3.6 for a summary.

But what if all does not go according to plan? What if MegaCorp is not happy with the job, or if some cluster members are unhappy with one (or more) of the others, or place the blame on one another for certain other failures? This is where a heuristic appeal to the Darwinian notion of "Survival of the Fittest" is made. Peer

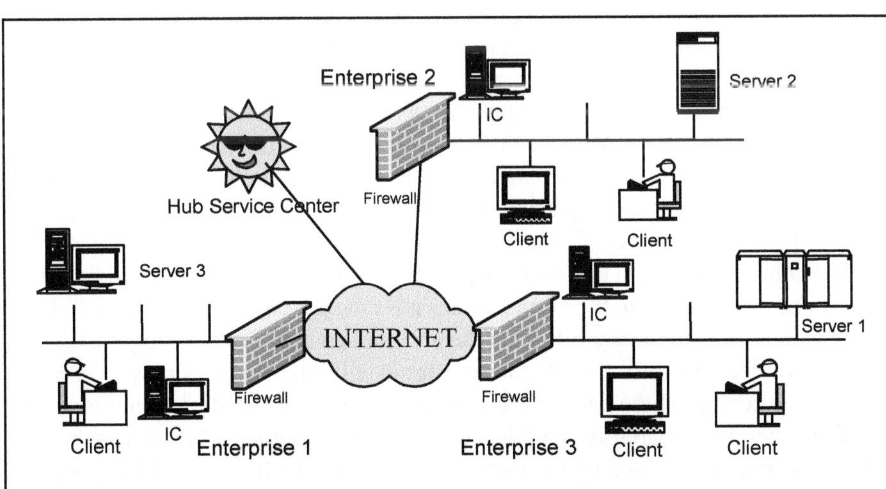

Figure 3.5. *Factory-on-Demand technology relations (adapted from Lee and Lau [1999])*

- MegaCorp advertizes order via hubs.
- SMEs read order and check listings to find compatible SMEs partners to team up with and form clusters.
- Clusters compete with each other in a bidding process to win the job from MegaCorp. Say cluster X wins and is comprised of 5 members.
- MegaCorp changes elements of its order request during the design and build phase. FOD allows this process, but one member of the cluster in particular has difficulties due to a lack of agility in responding to the changes. MegaCorp complains, but ultimately the project is seen through to a successful completion.
- Cluster X formalizes its documented peer review as done by each of the 5 members, which is then submitted to MegaCorp and advertized via the hubs. The weak member may deny its failures, but the feedback by the 4 other members and MegaCorp recognition of the problem make the facts hard to deny
- Cluster X members decide to maintain contacts, though the weak member is excluded. Subsequent positive reviews on other projects bolster the reputation of its remaining members, and further failures by the unresponsive SME cause is to be ostracised from further clusters and the firm is eventually forced to close.

Figure 3.6. *How a typical DNM setup works*

review works most effectively in a networked environment where large populations can lead to a high enough volume of reviews to hopefully quash those that are extreme in bias. Take, for example, Amazon.com or eBay with their review systems (Amazon reviews books while eBay reviews buyer and seller reputations). The peer review process, though almost entirely unprofessional, leads to the exposing and eventual weeding out of inferior products and people (i.e. bad books do not get purchased as frequently and unreliable buyers or sellers are effectively ostracized from the community). A similar peer review (including customer review) posted on a common messaging infrastructure via the appropriate FOD channel would serve a similar purpose. Ultimately, though working as a collective, the members of a cluster owe nothing to each other beyond the obligations of the current contract. Each is out to secure its own benefit presently and in the long run. Thus it is in every SME's favour to eliminate the competition of the others and, as in nature, it is the weakest that are first to go – this is natural selection as applied to the DNM world. A fundamentally important point to emphasize here is that in this world, SMEs are free to work with some, all, or none of their former cluster partners on future projects, though sometimes job constraints may limit the available choices. Clearly here (as in most networks), the more participants the better the efficiency, selection, and viability of the overall process.

Some clusters may feel that they are particularly well suited to a certain function, and may choose to maintain their connections after the job contract has expired. They are free to advertize their capability via the hub, and customers browsing the listings may see and decide to use a particular cluster as advertized. This may eventually lead to the outcome where certain clusters end up forming small conglomerates of their own, thereby apparently negating the DNM philosophy while still using the DMN resources as provided by the Factory-on-Demand technology. However, this does not pose much of an issue as nothing is to prevent other SMEs from forming clusters for a particular job and challenging the new mini-conglomerates, beating them at their own game. This, after all, was the whole purpose of DNM to begin with: giving power to the individually weak SMEs to challenge the large players on the global scale.

3.8 Summary and Conclusion

In the 20th century, our understanding of how production systems should operate was influenced by a number of key principles from Taylorism to Fordism to the Toyota Production System. In a global economy, the monolithic company has given way to a network of activities located in different countries. The advent of IT has further aided the creation of virtual enterprises: a community of factories each focused on what it does best. In this context, there is a tendency for smaller firms to evolve into loosely tied and decentralized federations or business units, making products and seeking alliance within and outside the "consortium" to serve customer demands to their best ability.

This chapter describes an attempt to develop a conceptual framework for a new form of production system that is unique from many perspectives. This new production system argues for a completely different form of SME collaboration from those already discussed in the literature. We introduced the notion of Dispersed Network Manufacturing (DNM), which advocates a reciprocal bonding among network members but calls for no obligatory egalitarian responsibility to each other. Our goal has been to provide an understanding of DNM, how it forms business ties, its characteristics, and why it works as a new business model. Building upon this, we then intended to discuss a possible realization of DNM, which we referred to as a Dispersed Manufacturing Network (DMN). The objective being to create a network of plants that are electronically linked so that the participating members, spread geographically, focus on their specialized tasks, yet also share their manufacturing and production resources to create a loosely structured and flexible enterprise.

Thc DNM World

- Less orderly and more dynamic
- Relationship is more important
- Structures form and dissolve as needed
- No fixed membership
- No one owns the network
- A protocol for communication between the units BUT not everyone has all the info (weak link of communication)
- Information openly distributed over a common channel where local clusters can form to competitively meet a need in the production chain.
- SMEs combine together to provide a complete production capacity that is sold to the customer
- Collaboration is a highly competitive process

The Old Paradigm

- More orderly
- Entity is more important
- Structures are to be maintained and optimized
- Information openly distributed but supply partners not free to act on it as they wish
- SMEs are contracted on an individual basis and their products are combined by the customer
- Collaboration is a highly controlled process

Figure 3.7. *Comparison between the DNM World and the old manufacturing paradigm*

In the DNM universe, because of the network's requirement to re-form itself to the needs of each unique incoming project, SMEs have the ability to rapidly develop and enhance their internal production capabilities. Each new incoming project offers the chance to reaffirm those processes that are highly effective while discarding those that are deemed ill-suited. In short, such dynamic and adaptive networks are believed to have competitive advantages on both the collaborative and individual scales.

DNM and DMN are arguably powerful concepts that are evolving into a new production paradigm beyond what we have previously witnessed. It is likely that many companies and many managers have some intuitive grasp of the DNM world and the opportunities provided by forming a DMN. Few, however, might understand them thoroughly. It is therefore believed that this paper has contributed to a better understanding of these concepts and the principles involved for their existence (see also Figure 3.7 for a comparison of the manufacturing paradigm based on the DNM business model with the traditional one).

The DNM world is still in its infancy and many interesting and challenging questions have yet to be investigated empirically or otherwise.

Appendix: Can DNM Prosper Among "Selfish" Individually Small Companies?

There is a common-sense feeling that win–win cooperation systems are unstable and difficult to achieve in the long run, despite the benefits of cooperation. Even in situations where everyone is better off together than they would be apart, there seem to be those who think they deserve a "larger slice of the cake" than they are getting, disturbing the stability of the cooperative effort. Individual interest often conflicts with collective interest. Adding to that, if there is not a way to ensure that the benefits of cooperation are going to be restricted to those who actively cooperate, free-riding and cheating start taking place, jeopardizing the position of those who play by the rules.

That being said, should we still consider it reasonable that companies motivated by their own self interest – like the ones our market oriented society generates – cooperate with partners in the long run, or rather consider the notion romantic and naive? That question is relevant to managers, because in order to decide on multi-company arrangements, like the manufacturing consortia, or the Japanese Keiretsu, companies have to take the expected future actions of their partners into consideration as they develop their own long term strategic plans. The more they feel comfortable to commit to their relationships in the long run, the more they will accept having their systems and processes integrated with one another and the more freely they will allow information to flow and be shared along the supply chain. The more processes are integrated and information is shared, the more synergy results from cooperation.

Economists have gone far in trying to understand the problems of conflicting interests at the individual and collective levels. The game theory they have developed shows that such conflict is due to a rational tendency of maximizing individual gains at the expense of the group's interest, especially when there is not an effective incentive/punishment policy applied to ensure that no one tries to cheat the system. The only way to get successful cooperation among people, according to economics is to make cooperation profitable for each and every co-operator, individually. This means avoiding the "tragedy of the commons" among partners, which is an issue that has to be systematically addressed.

Biologists have also had a lot to say about cooperation. One of the first things they discovered was that natural selection is about the strongest individuals prevailing, rather than the strongest group of individuals or species. Therefore, no matter how much a feature benefits the group or the species, it will not evolve unless it also benefits the individual. That is so because, in a world of "selfishness", a feature that makes an individual help others without retaining any self-benefit would die out, even if the benefit provided to others far out-weighed the costs assumed by the individual itself. So, for cooperation to evolve, ways would have to be found of building cooperative organizations out of self-interested components, which, in other words, means it would have to be made in the interest of individuals to cooperate.

That seems to pose a great challenge for the evolution of cooperation, regardless of whether we are talking about molecular processes, cells, animals, humans or organizations of humans, like intra or inter-company relationships, like supplier-buyer relationships.

But, in fact, biologists found much evidence in nature that cooperation can indeed happen among "selfish" individuals, once incentive/punishment mechanisms are established to stimulate cooperation and avoid free-riding and cheating. Such mechanisms are essential for, if they are not in place, cooperation does not lead to a better performance of the more cooperative individuals who are out-competed by less altruistic individuals in their struggle for Darwinian survival. The eventual result is that when co-operators die out, so does cooperation.

Biologists have found several examples of mechanisms that support cooperation between "selfish" multi-cellular animals. Some of them cooperate only with other members of the family, who they expect to, most likely, return their favours. This seems reasonable as, family members have a greater probability of sharing the same co-operator gene. Other animals rely on what has been called reciprocal altruism. This is based on the fact that an individual does not get into disadvantage if it helps another, provided the other helps back in a future situation. Of course for cooperation based on reciprocity to work, there has to be a fair balance between the favours being exchanged, otherwise, those who do more for their partners than they get in return are out-competed. Only clever animals can keep a good record of other individuals who are good co-operators and of those who are not, in order to avoid being cheated and out-competed in their relationships. Scientists have discovered that bats, elephants and a few other animals are capable of developing reciprocal altruism relationships. Humans can, of course, also do that, at least in groups that are not too large.

If we think about business, we notice that people, and companies, tend to cooperate with others they know to have a reputation as good co-operators. At the same time, they are reluctant when they have to relate to strangers, especially when no future cooperative opportunities are foreseen. The theory of games shows that good collaborators rapidly turn into cheaters when a game gets closer to the end and participants do not feel that cooperating is the best approach anymore in order to succeed.

In a scenario of local markets, individuals and, less usually, companies could leave a bad reputation behind and find other partners elsewhere that were unaware of their previous "bad behaviour". But that attitude is not a wise one in a scenario of a global market with good information diffusion. So, the need to keep a good reputation in order to be accepted in future cooperative arrangements is a relevant mechanism to support the establishment of cooperation among self-interested organizations.

We have been hearing a lot about supply chain management in the last few years, and the ways in which competition happens at the value chain level and not at the business unit level, but many still consider it a romantic concept. They like the idea until it comes to the point they realize the only way of achieving its full potential is through sharing information, resources and a common strategy with the other members of the value chain. Then they start thinking about the possibility of cheaters and free-riders to flourish throughout the group and the tragedy of the commons takes place: if someone is going to get benefits without putting the effort in, why not let it be me? Or, at least, I am not going to be the fool to carry the others around on my shoulders! And there is, unfortunately, no solution to that cruel situation in which people know they would do better if they only cooperated, but they will not because there is no guarantee that the others will, that is unless mechanisms are put in place to prevent cooperation from failing despite everyone's "selfish" nature.

Reciprocal altruism reliant on good reputation is one such mechanism that helps to establish an environment of trust among selfish partners. Such trust should not be naïvely based on the belief that the cooperating partners primarily seek the benefit and the survival of the group, though, because they simply do not. They first think of their own survival and success. Altruism becomes only part of our business behaviour when it is that kind of reciprocal altruism, discovered by the biologists, that is based on the likelihood that the favours returned, preferably, outweighing the ones provided. In our market society, companies are expected to be selfish, because they were created to generate profit for their shareholders, not for the shareholders of their partners, unless they can conciliate the two. Companies are committed to their own shareholders and any other commitment they make is expected to be in their interest. So they are going to be selfish in their behaviour and will only cooperate with others to the extent it is convenient from their point of view. That means they will cooperate with partners while the cooperation takes them to where their stakeholders consider appropriate for the company's success and survival in the market. That may not sound as the best environment in which for cooperation to flourish, at least at first glance, and it is understandable why so many people are sceptical about the possible work-outs of DNM. But, cooperation may end up being the best way for "selfish" companies to accomplish their goals. As biologists have shown us with their examples from nature, if companies are able to formulate the proper mechanisms to manage cooperation, providing for a reasonable distribution of the benefits achieved from cooperation among the partners (that means avoiding cheating and free-riding), they will be able to better fulfil their own selfish goals. Co-operation can provide firms with a very good edge in the market, for it makes them much more flexible and agile in approaching the market's needs, as has been discussed throughout this chapter.

The bottom-line for a cooperation arrangement to be worthwhile and merit attempt is that the sum of the parts must be greater than the whole. In other words, collaboration should maximize combined capabilities and enable each party to realize its strategic goals, while allowing for the partners' goals to be also accomplished and while providing integrated solutions to the end customers' needs. If collaboration can lead to a situation where 1+1=3, success will only be a matter of figuring out how to share the extra 1 they get, without making the selfish partners feel they are getting less than they should out of the arrangement.

Acknowledgements

The Appendix is based on the earlier contribution of Alex Graeml during the visit of Hamid Noori to Haas School of Business, UC-Berkeley. The contribution of Arshia Noori, who is currently completing his post-graduates degree at UBC is acknowledged. This research was partially supported by Laurier Chair in Enterprise Integration and Technology Management, and the Hong Kong PolyU under the Distinguished Visiting Scholar Scheme. The authors greatly appreciate the constructive suggestions made by Rob Dekkers and the two anonymous referees.

References

Applegate, L.M. 1999. "Time for the Big Small Company". Mastering Information Management, Financial Times: 1–12.

Biggiero, L. 1999. "Market, Hierarchies, Networks, Districts: A Cybernetic Approach". Human Systems Management, 18(2): 71–86.

Dekkers, R. 2005. (R)Evolution, Organizations and the Dynamics of the Environment. Springer, New York.

Dyer, J., Cho, D.S. and Chu, W. 1998. "Strategic Supplier Segmentation: The Next Best Practice in Supply Chain Management". California Management Review, 40(2):57–77.

Enright. M., Scott, E.E. and Chang, K. 2005. The Greater Pearl River Delta and the Rise of China. Wiley, New York.

Hannan, M. and Freeman, J. 1993. Organizational Ecology. Harvard Business School Press, Boston, MA.

Hanson, P. and Voss, C. 1999. "Taylor to Toyota to Technology". Manufacturing Engineering, February: 11–14.

Haeckel, S.H. 1999. Adaptive Enterprise-Creating and Leading Sense-and-Response Organizations. Harvard Business School Press, Boston, MA.

Holland, J.H. 1995. Hidden Order: How Adaptation Builds Complexity. Addison-Wesley, Reading, MA.

Kauffman, S.A. 1994. "Whispers from Carnot". In Complexity: Metaphors, Models and Reality, Cowan, G.A., Pines, D. and Meltzer D. (Eds.). Perseus Books, Cambridge, MA.

Lee, W.B. and Lau, H. 1999. "Multi-agent Modeling of Dispersed Manufacturing Network". Expert System with Applications, 16(3): 297–306.

Li, L., Alderson, D., Tanaka, R., Doyle, J. C. and Willinger, W. 2004. "Towards a Theory of Scale-Free Graphs: Definition, Properties, and Implications." Technical Report CIT-CDS-04-006, Engineering & Applied Sciences Division, California Institute of Technology, Pasadena, CA.

Magretta, J. 1998. "Fast Global Entrepreneurial: Supply Chain Management, Hong Kong Style: An Interview with Victor Fung". Harvard Business Review, 76(5): 103–114.

McKelvey, B. 1999. "Avoiding Complexity Catastrophe in Coevolutionary Pockets: Strategies for Rugged Landscapes". Organization Science, 10(3): 294–321.

Montreuil, B., Frayret, J.M. and D'Amours, S. 2000. "A Strategic Framework for Networked Manufacturing". Computers in Industry, 42(2–3): 299–317.

Moran R.T. and Riesenberger, J.R. 1994. The Global Challenge: Building the New Worldwide Enterprise, McGraw Hill, New York.

Newman, M.E.J. 2003. "The Structure and Function of Complex Networks". SIAM Review, 45(2): 167–256.

Noori, H. 1998. "The Transition from Low-Valued Repetitive Manufacturing to Technology Hubs: The Influence of Globally Operating Companies". The Journal of High Technology Management Research, 9(1): 69–86.

Noori, H. and Lee, W.B. 2002. "Factory on Demand and Smart Supply Chain". International Journal of Manufacturing and Technology Management, 4(5): 372–383.

Noori, H. and Mavaddat, F. 1998. "Enterprise Integration: Issues and Methods". International Journal of Production Research, 36(8): 2083–2097.

Noori, H. and Radford, R. 1995. Production and Operations Management: Total Quality and Responsiveness. McGraw Hill, New York.

O'Neill, H. and Sackett, P. 1994. "The Extended Manufacturing Enterprise Paradigm". Management Decision, 32(8): 42–49.

Pietarinen, A.V. 2004. "Multi-Agent Systems and Game Theory - A Peircean Manifesto". International Journal of General Systems, 33(4): 395–414.

Pine, J. 1992. Mass Customization: The New Frontier, Harvard Business School Press, Boston, MA.

Porter, M. E. 1998. "Clusters and the New Economics of Competition". Harvard Business Review, 76(6): 77–90.

Robertson P. L. and Langlois, R. N. 1995. "Innovation, Networks, and Virtual Integration". Research Policy, 24(4): 543–562.

Schilling, M. A. and Phelps, C. C. 2007. "Interfirm Collaboration Networks: the Impact of Large-Scale Network Structure on Firm Innovation". Management Science, 53(7): 1113–1126.

Shaw, T. E. and Lengyel, A. (1996). "Systems Analysis of the Agility Imperative". In: Proceedings of Fifth National Agility Conference. The Agility Forum, Lehigh University.

4 Self-Similarity and Criticality in Dispersed Manufacturing: A Contribution to Production Networks Control

Hermann Kühnle
Otto-von-Guericke-University of Magdeburg

Abstract Networked organisations, like those in Distributed Manufacturing, are structures that fundamentally differ from hierarchical organisations, as they emphasise speed, re-linking and reconfiguration. For the control of these networks, alternative procedures, derived from a newly introduced criticality framework, are outlined. For quick and effective realignment of units and the reallocations of resources, the self-similarity principle proves to be useful for the description, control and evolution of manufacturing networks. Moreover, pilot implementations of these elements, as described, allow expectations of efficient practical solutions. At the task level, it has already been possible to cast highly adaptable control set-ups, based on the same philosophy, into an industry pre-standard. This chapter extends these solutions to the more sophisticated network control on the enterprise units' level and is intended to contribute to the development of a new generation of ERP systems, covering also Dispersed Manufacturing.

Keywords Control decision, Critical unit, Footprint, Pattern similarity

4.1 Introduction

Global presence in many regions and markets is essential for suppliers and manufacturers as well (Dekkers and van Luttervelt, 2006). This trend has caused many companies to expand across borders and has extended manufacturing networks. Particularly, companies experiencing rapid international expansion through mergers and acquisitions are suddenly faced with the challenge of effectively structuring, managing and operating a network of geographically dispersed factories with world-wide transfer of assembly and manufacturing operations between multiple production sites, even for similar products, in different countries (Clarke, 2005). Structurally, organisational units must continuously be added or shifted to a different location, enforced by market demands, strong competition or high innovation pressure, and not just as a one-time result of organically developed growth.

A very general description for Dispersed Manufacturing Networks can be given as "organisational manifestations for collaboration and coordination across loosely connected agents" (e-Volution II, 2004). Such Dispersed Manufacturing Networks comprise the total primary process, product development and supply chain, in an international setting, relying on applications of information technology and data-communication to exchange information and to co-ordinate actions.

The inevitable consequence: management of production becomes complex and more difficult. While the network characteristics, such as breaking up, re-linking or re-configuration of important manufacturing structures, have become decisive management issues (Cooper *et al.*, 2006), theory and practice still focus on approaches and solutions simply operating ICT connected units in rather rigid supply chains. Consequently, the management techniques and ERP procedures for production systems planning, decision and control procedures, developed for hierarchical organisations, are still applied nearly unchanged for inter-enterprise network organisations. The computational results for static organisations inevitably suffer from methodical shortcomings due their one time one place design. The planning and control results show all characteristics of periodic interventions, based on elaborate top-down procedures that do not follow the continuous changes that are taking place in complex company networks (McKelvey, 2004).

Dispersed Manufacturing Networks, however, are versatile, dynamic and self-organising structures that require a different kind of decision support not available thus far (Dekkers, 2005). On the other hand, methods and instruments for prioritising control decisions and the quick implementation of efficient processes in networks is becoming increasingly important to practitioners. There is a strong agreement on deficits and fields of action (Bennett and Dekkers, 2005); research has been carried out in various areas, using varying terminologies, such as Virtual Organisation (VO), Virtual Enterprise (VE), Extended Enterprise (E2) (Davis and Spekman, 2004), Concurrent Enterprising (CE), or Collaborative Network Organisations (Camarinha-Matos and Afsamanesh, 2005). Extensive research has been done in the fields of dynamic and adaptive capabilities (Gunasekaran and Yusuf, 2002; Srai and Gregory, 2005). All work and terminology hint at profound needs to initiate additional efforts aiming at new network theories, based on interdisciplinary approaches, such as information sciences, complexity theory, topology and organisation theory (Kühnle, 2007).

4.2 Units in Manufacturing Networks

Dispersed Manufacturing may be seen as a new pattern of inter-firm relationships evolving into a different form of enterprise-wide integration of inter-entity processes. If production systems "disperse" their value chains, engaging a large number of smaller units all over the globe, the value created increasingly appears as a result of geographically distributed networked operations and services, representing in total at least the sum of all necessary resources. As the responsibilities for operations are tied to organisational units and their socio-technical nature, Dispersed Manufacturing shows all features of human influenced complex networks' building (e.g. trust, individual preferences), as well as planning and execution of efficient processes within networks exceeding the scope of ICT with repetitive process routines and standardised functionalities.

Collaborative entities will increasingly turn interoperable sets of units into densely interconnected networks, as volatile market conditions and quick changes in configurations places comprehensive communication into a key role (Figure 4.1). Not only do the singular units and their behaviour change but their interrelations and cross impacts are decisive factors of success, posing challenges to the management (Citrin and Neff, 2000) to bring in other qualities of leadership (such as clear strategy, building flat organisations, managing via business models, standardising methods, etc.) as well as to develop better linking capabilities (Teece *et al.*, 1997). As all complex networks, Dispersed Manufacturing includes the following attributes:

- they consist of interacting units;
- the units may vary and change over time;
- the units have varying degrees of internal autonomy;
- the units are coupled in a non-linear fashion;
- behavioural patterns are created through myriads of interactions;
- collective behaviours emerge beyond the control of any single firm;
- networks create and destroy transitory traditional systems, as reconfiguration is adapted.

Resource allocation and scheduling has to be totally re-evaluated with respect to this new context. Since networks cannot rely upon central control, adequate substitutes will have to be established by the entities' interactions. Whereas the total network configuration level (GERAM, 1997) calls for game theory applications (e.g. for units evaluation, network access), the network control level emphasises negotiations between entities concerning parameters and objectives. Therefore, the network's configuration is primarily driven by local decisions of entities and their direct communication. All entities will have to "control" their interrelationships by applying decentralised autonomous procedures.

Within this context, the entities make use of their own preferences, information structures and evaluation patterns. Evidently, prescribed and objective-led preferences

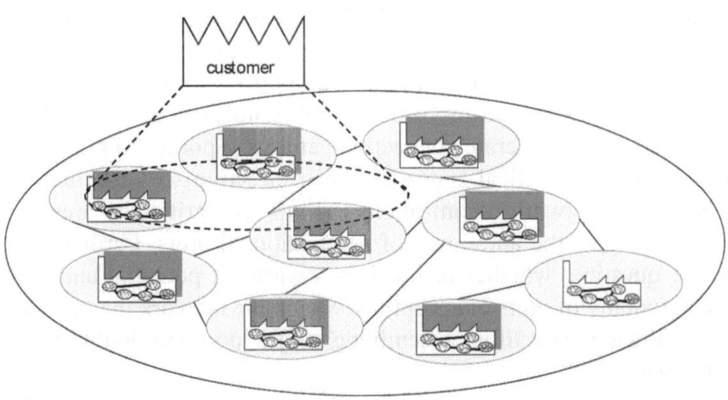

Figure 4.1. *Network of units*

for the decision making within units have to support the networks' strategy and ensure the network interests. Under the assumption of rationality, the resulting entities' decisions for the entire networks may be esteemed as fairly good. In any case, the units have to support the Dispersed Manufacturing networks' objectives in all situations and be in line with ever-changing configurations. This suggests that while utilising the advantages of the units' autonomy, the network can neither tolerate the units' complete independence nor its underperformance.

The implementation of adaptation-based decision structures (Ivanov *et al.*, 2006) is one way to manage Dispersed Manufacturing, ensuring adequate degrees of co-operation as well as capabilities' engagements to satisfy the networks' objectives. They include parametric adjustments and alignment as well as resetting of the objectives and complete network reconfigurations. One important instrument, which has been applied very successfully for the control of adaptive structures, will be introduced subsequently.

4.3 The Autonomy and Criticality Framework

Effective network control decisions have to optimise the network units in economising resources, fulfilling objectives and strengthening the network on the whole. If one or more units do not offer the required capabilities and support to the network adequately, these units may become "critical units", or more precisely are considered as units moving into so called critical states, i.e. their role within the network must be evaluated.

In an organisational context critical states have not been elaborated on so far. These concepts for decision models and criticality thinking have been applied in other areas, such as distributed query composition, workflow composition and Hierarchical Task Network (HTN) planning where agent technology has been applied successfully (Bratoukhine *et al.*, 2003; Deen, 2003; Budenske *et al.*, 1998; Peng *et al.*, 1998; Jennings and Wooldridge, 1998; van Brussels *et al.*, 1998). Furthermore, it has been argued that it is particularly suitable for integrating bioinformatics resources (Karasavvas *et al.*, 2005, Merelli *et al.*, 2007). Fractal supply networks (Kuehnle, 2005) are just one new application field of complexity theory (Olson *et al.*, 2001; Barbasi, 2005). Also, connection patterns of the cerebral cortex consist of pathways linking neuronal populations from whole brain regions to local mini-columns that may be successfully examined by measures of complexity and described by self-similarity procedures to generate connection patterns (Sporns, 2006).

As this concept of critical states has been effective for other networks, it is bound to be apt for network organisations as well. The criticality approach allows adequate modelling of the autonomy of organisational units. Furthermore it helps to answer the question, whether or not it is a waste of potential and resources to tolerate the autonomy of units or accepting units in the network that are (repeatedly) unable to avoid such criticalities resembling single loop and double loop learning (e.g. Senge, 1990).

The proposed network control procedure, therefore, links the degrees of autonomy of units to their abilities to avoid critical situations. This way of modelling and coupling the criticality and units' autonomy has proven to be a powerful base for the decision support in network control.

In the context of Dispersed Manufacturing Networks (e.g. see Noori and Lee, 2006), because they correspond to successful approaches (Klostermeyer, 2002) in distributed manufacturing control, Spaces of Activity (SoAs) may be introduced as a measure to monitor the state variables that map the individual units' activities and success. It is set up of the units' objectives (output of the network strategy) mapped onto one dimension, with the available resources and constraints represented by the other two dimensions of the Space of Activity. For decisions, whether or not interventions are executed, we envision this Space of Activity as a criticality threshold, to immediately initiate actions, when overriding the interaction flag, i.e. the units' observable position is outside of the SoA, so the entities' criticalities exceed the assigned thresholds (Figure 4.2). The initiated actions will try to ensure the fulfilment of the objectives of the entities so the total networks' objectives are met.

Usually there is more than one way to interpret a particular critical situation i.e. the available solutions may be changes in responsibilities, the adjustment of autonomy, adaptation or initiation of network restructuring. Differentiated measures for the criticality of situations are therefore useful for pruning adequate solution spaces for the arising problems in the network control.

Comparison integers may be introduced for characterising repeated not admitted position situations that indicate the least number of comparisons during a particular decision situation. The higher the number of queries characterising a specific "not admitted" decision situation, the more critical that situation becomes. If the number of comparisons, resulting in "not admitted positions", is higher than the assigned integer, it will run into a "more severe" decision cycle. Eventually the unit's SoA positions result in decisions on maintaining the self-organisation mode, reducing or removing the autonomy and calling for PN interference. In criticality terms, each unit may do the following

i. Decide on appropriate methods, tools, etc. in order to achieve the objectives negotiated and agreed upon. Units remaining within the admitted SoA are allowed to execute autonomous decisions. Prerequisites are resources, e.g. budgets, competencies, technical and personnel availability, and constraints such an unit may have to face, e.g. legal restrictions and capacity limits.

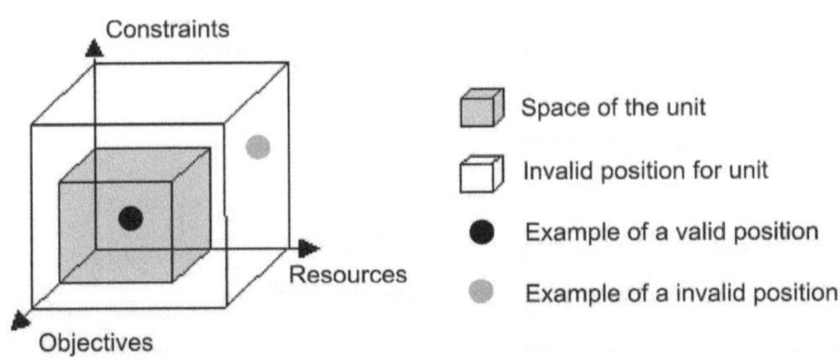

Figure 4.2. *Space of Activity (SoA) – dimensions and criticality threshold of units' – viewing the actual valid or invalid positions of units in the network*

ii. Lose its autonomy, if non-critical positions within the unit's SoA are not achieved by its own efforts. Instantly, network mechanisms are activated preventing the deviations and providing for the achievement of the overall network plan.
iii. Be replaced by new or other network units and be removed from the network if ii is repeatedly experienced.

Dependant on the unit's (in)ability to cope with changes in the environment, network order parameters may gain influence on the units' activities according to the subsequent scheme:

A. Admitted position: no action.
B. Sporadically not admitted position: non-critical, self-organised optimisation by the unit is demanded;
C. Repeatedly not admitted position (within threshold): (critical) autonomous self-organisation, where the critical state is overcome by the respective unit.
D. Repeatedly not admitted position (exceeding threshold): (critical) interaction, where the network asks for changes in criticality values (Space of Activity Volume), while presenting expected benefits/drawbacks that account for the critical situation.
E. Repeatedly not admitted position (exceeding the threshold by far): (critical) restructuring, where alternative structures (breaking up of links, generation of new interconnections, and introduction of new entities) are checked and the results are compared.

In addition, once executed, the framework defines a network request, i.e. an explanation (global explanation) for the choices made during the decision (see Figure 4.3). For economic reasons the network will try to keep the volumes of the Spaces of Activities somehow limited. The production network may react on any increase of market complexity (diversity, uncertainty and unpredictability) by expansion of the SoAs affected (if affordable). More foreseeable steady conditions allow shrinking the SoAs' volumes.

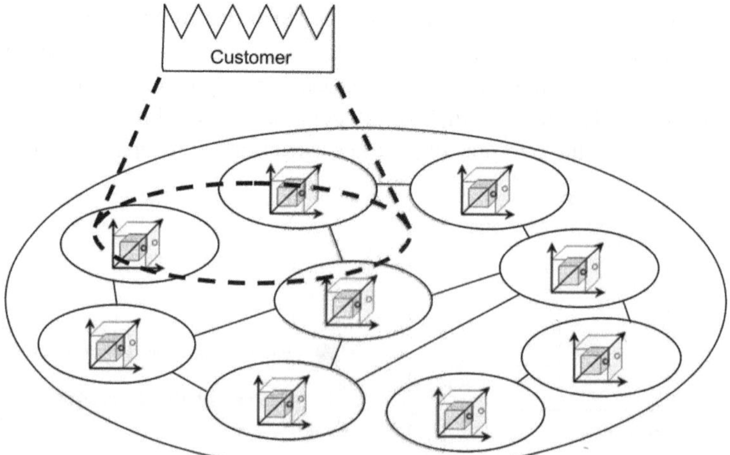

Figure 4.3. *Dispersed network example as network of Spaces of Activity*

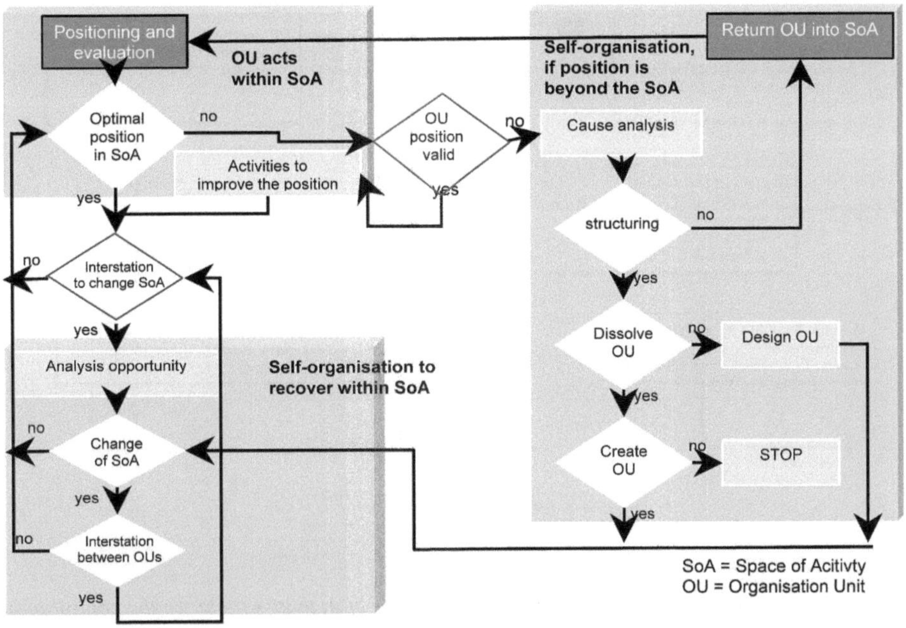

Figure 4.4. *Total network control loop for self-optimisation of units for short term re-configuration of Dispersed Manufacturing Networks*

The proposed planning and control procedure has been verified in Dispersed Manufacturing Networks. Plans, assignments, units, responsibilities, etc. are continuously rearranged, new processes established, broken up or reconfigured. The control procedure uses intensive communication as well as objective driven adaptations and configurations.

From the viewpoint of the units, splitting, removal or re-linking of units are decisions to be taken in cases of non-admitted positions and if the deviation cannot be avoided by the unit's very own efforts (Figure 4.4). Medium and long-term network access and control procedures are assumed to be overlaying (e.g. as specified by GERAM).

The criticality framework as outlined has proven to be especially helpful for network units to make the most of the available facilities while increasing their trust in the network. The implementation of the framework improved the inventory levels and the costumer service performance in a balanced manner across the network.

The criticality concept helped the network members to better understand how the network control arrived at a particular decision.

An example of a network of companies that manufactures tools and abrasives, may illustrate this control logic set up. The Space of Activity concept was implemented in order to improve the adaptability and the production flows as described. The procedure has been implemented under ERP, to gradually take over the control functions (Figure 4.5). Simple agent-based track and trace functions were the first additional features for monitoring the units' positions. Other test case improvements were measured as in Table 4.1

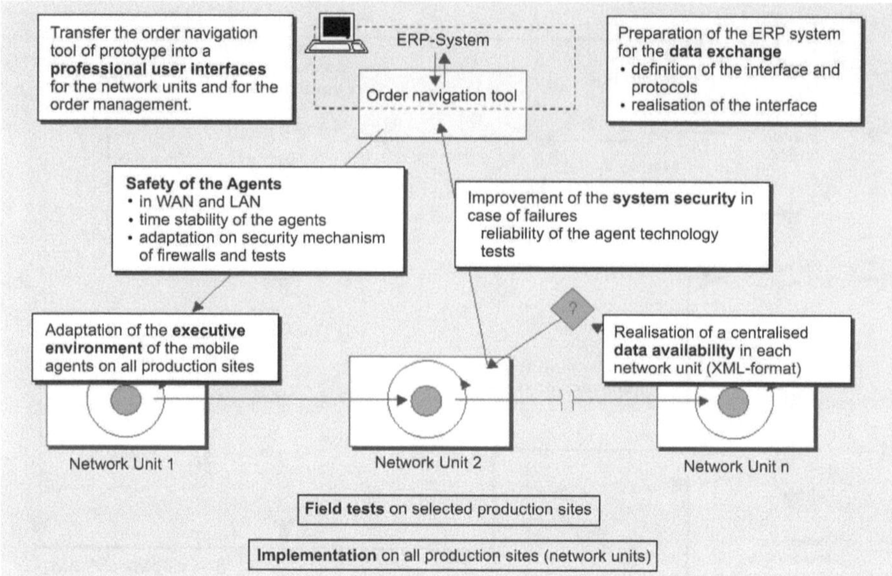

Figure 4.5. *Steps for the implementation of the Space of Activity's logic for the control of a Dispersed Manufacturing Network*

The criticality control mode, as outlined, differs from traditional planning and control, where elaborate plans are calculated for each unit, covering discrete planning rhythms and horizons. It is continuous, distributed and concurrent, generating solutions in a gradual manner. In the test cases central control functions of ERP had to be maintained to ensure the fulfilment of plans that had already been generated on the base of cyclic procedures. Although it was known that these plans are incorrect instantly after their set ups, the existing ERP installations had to stay connected, as momentarily no software solution package is available that incorporates the alternative logic. However, some leading software vendors are intensively working on the "next generation" of products that will include procedures of this kind.

4.4 Self-Similarity to Cope with Network Complexity

A key property of all networks is complexity (Kauffmann, 1993) with particular impacts on organisations that are of network nature (Dooley and Van de Ven, 1999). Dispersed Manufacturing Networks are re-configurable dynamic organisational set-ups, inter-linking or detaching units as well as continuously establishing and optimising versatile process chains. Structures "emerge" as results of local interactions as the Dispersed Manufacturing Networks configure and self-organise towards market opportunities. It appears that a few configurations are more favourable than others in some way. Each network reconfiguration step raises the question of how the dependencies and the simultaneous planning actions have influence on the preferences, the attributes and the objective fulfilment of the involved units. On

Table 4.1. *Effects of the implementation on performance indicators*

	Starting situation	Objectives		Results		
Lead time	10 min	↓	20%-30%	↓	10% - 22%	decrease
Inventory	60% *	↓	40%	↓	48%	reduction
Deliver ratio	90%	↑	98% -100%	↑	95%	increase
Productivity	80%	↑	90%	↑	82%	increase
Overhead	85%	↓	50%	↓	68%	decrease

the other hand, changes at all levels of detail of any unit may influence the entire network.

Complexity is always referred to as being rather difficult to be mastered, so adequate simplifications are admitted. For the control of the Dispersed Manufacturing Network, instruments that support quick and exhaustive overviews as well as efficient criticality decisions and interventions are adequate. As decisions and interventions might be necessary for all units on all levels of detail, there will be large numbers of entities and relationships to be monitored and managed within a network. Therefore the models, used for the units, have to be both: rich and simple. According to first pilot cases (Kühnle and Wagenhaus, 2004), the SoA models of the units prove to be powerful enough to cover the network control problems adequately as long as fundamental principles for complexity reduction, as described below, are in place.

Throughout history, metaphors and analogies have often been successfully used to describe concepts on current technologies. Most metaphors and analogies were helpful because they had extended the knowledge acquired by the scientific and the technological developments to other areas, illuminating them from a novel perspective and making it simpler to understand and to handle. Examples are the description of a factory as the body (spine concept) or the description of the brain as a computer.

For manufacturing networks and their attributes, it has been fruitful to take into account analogies from geometry, as it shows that very complex objects can be characterised by simple recursive algorithms (complex nature – simple code). Mandelbrot named an important kind of complex geometric structures (fractals), emphasising their fragmented character (Mandelbrot, 1982). A fractal may be a fragmented geometric shape (Figure 4.6) that can be subdivided into parts, each of

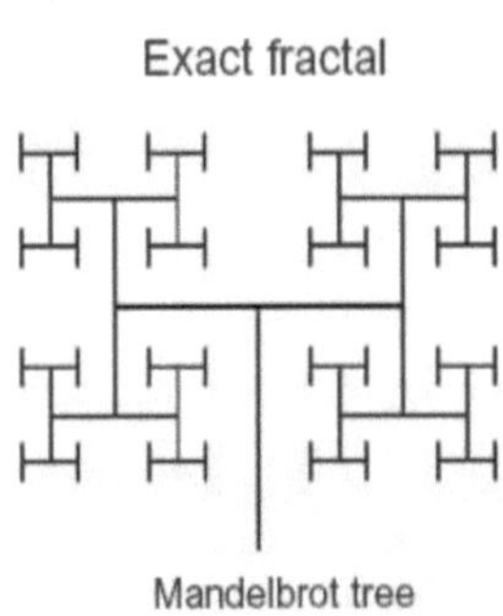

Figure 4.6. *Mandelbrot tree: Exact fractal – an example for a self-similar structure*

which is (at least approximately) a reduced-size copy of the whole, and is therefore called self-similar.

Especially for networks in manufacturing, such as Dispersed Manufacturing Networks, this nature of "metaphors" appears most promising as it synthesises systemic thinking (subsystems, partial systems [Klir, 1969]) and important complexity attributes (e.g. pattern cognition, diversity [Webb *et al.*, 2004]). Fractal entities, envisioned as parts of organisations that self-organise, self-optimise, are linked through intensive information and communication and strive for the given objectives (Warnecke, 1997). Following this thought an independently operating network unit may be interpreted as a fractal or a fractal entity, described by its objectives and resources, representing – with respect to these attributes – a self-similar "copy" of other network entities as well as of the total network.

4.4.1 Extension and Breakdown of SoA Objectives' Axes

For the continuous update of all objectives and resources in a quick and easy manner, predefined patterns (also self-similar) have proven to be a valuable support that just need to be modified according to the emerging constellations. In order to detail the objectives on the network level as well as on the unit level, a useful pattern may be defined by a standard objective list where the priorities are to be defined and network specific parameters may be inserted.

The fulfilment of the objectives within any configuration does not necessarily imply that those goals will be identically supported by all assigned lower-level goals. It is the intensive communication between the levels and the sub-levels that ensures the intended self-similarity as well as the resulting consistency of all objectives.

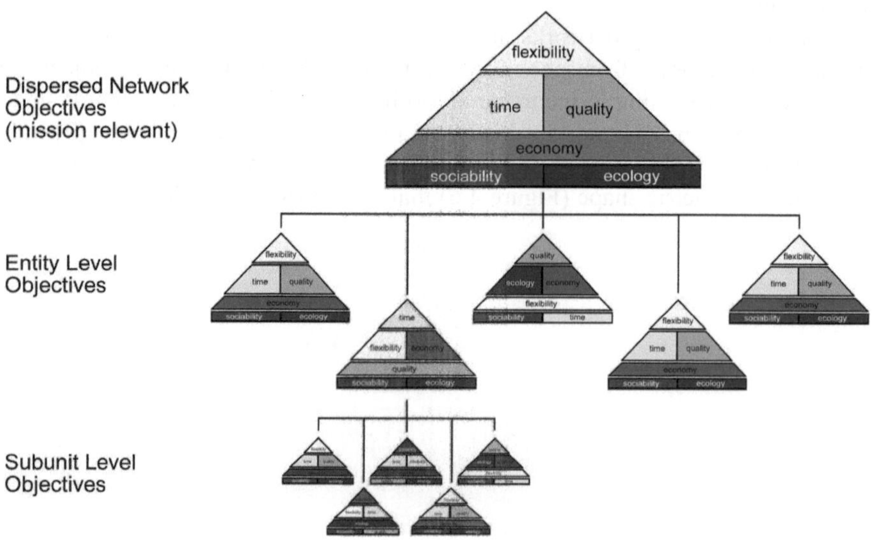

Figure 4.7. *Self-similar objective break down objective pattern*

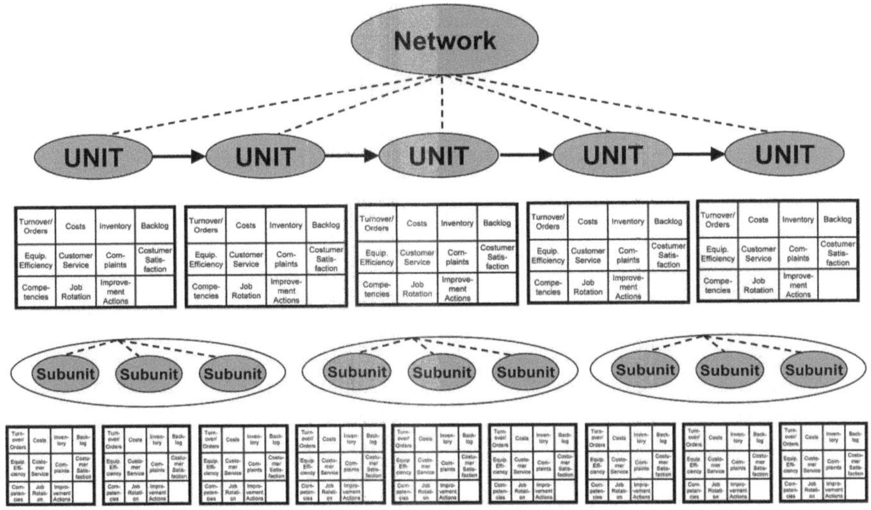

Figure 4.8. *Performance indicators' breakdown for a network of switch makers*

In most practical cases there are the (standard) objectives: time, quality, speed and cost to be weighed according to strategic orientations of the networks and the units. The units' objectives/goals and its priorities result from the network's strategy, broken down (self-similarly) into the units and subunits (Figure 4.7).

With regard to both basic levels, objectives and their combination as well as their consistency can be visualised by using self-similarity and the goal pyramid.

The breakdown of performance indicators may be interpreted as as well a self-similarity outcome, as the indicators' set for a manufacturing network for electrical switches (Figure 4.8) shows.

4.4.2 Extension and Breakdown of SoA Resource-Layers for Production System Specification

For quick descriptions of the resources' situation, the constituent components may be decomposed, so the configuration of processes as well as the evaluation of performance indicators on any level of detail will be supported. Aspect-wise systemic decompositions have been successfully applied, distinguishing between information flows, organisation and processes, such as the specification of the CIM/OSA framework (Kosanke and Zell, 2005) and consecutive standards. In self-similarity studies of fractal organisations, specific decompositions have been useful, subdividing into six layers (culture, strategy, socio-informal layer, finance, information, and processes) where the last four layers may be regarded as the resources' scheme (Kühnle, 2005) (Figure 4.9). Equivalent resource co-ordination schemes for networked production have been proposed elsewhere (e.g. 1. physical goods, 2. information, 3. people and 4. finances [Barlett and Goshal, 2000]).

The layers self-similarly replicate within the (sub)units as well as on the corporate or network levels. For better support of the communication, agreements

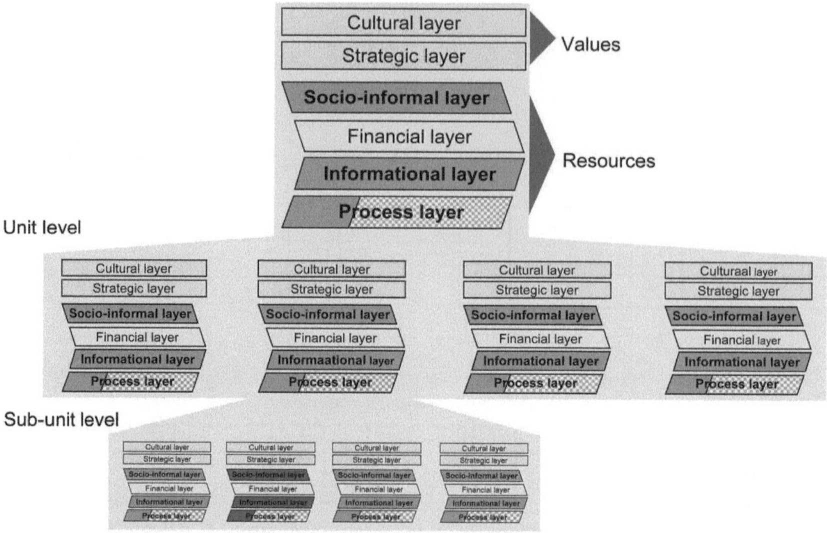

Figure 4.9. *Self-similarity principle applied to the enterprise aspect layer decomposition (Kühnle 2005) with focus on processes, information and organisations*

on IT platforms, on cultural values as well as on standards for interoperability are most favourable.

In order to support the management of a Dispersed Manufacturing Network, simple patterns of resources' structures may be prescribed internally for mandatory use. The key elements most frequently referred to are:

- company principles, (e.g. perfect quality, standardisation);
- level of detail invariant KPIs (e.g. time, value);
- management methods, (e.g. 5s, Kaizen, VSD).

The aspect layers, as described above, are useful to further classify the defined network-specific (self-similar) patterns that may be interpreted as a part of the general definition of a company's specific Dispersed Manufacturing System "standard".

Such standards signalise to all entities as well as to each network partner, the preferred methods and indicators. What the expectations of the network are and how potential partners increase their chances of eligibility and support to become part of this specific Dispersed Manufacturing structure are clearly stated. Many companies visibly highlight the importance of such standards by showing their commitment to these production systems. e.g. through identification of their company (or brand) names (Toyota, Bosch, Audi, GKN, etc.). Often, all vital elements of these company standards are visualised by self-explanatory symbols as hierarchical trees or mind maps, being referred to as the companies' "footprints" – an excellent metaphor for the self-similarity in the context of Dispersed Manufacturing itself (Table 4.2).

These examples demonstrate that managers are well aware that network control requires procedures that are different from what they are actually using in hierarchical organisations. The permanent need to restructure and re-link obviously brings about successful practical solutions that strongly involve the principles of complexity. However, the focus of such control efforts is still on the design of units'

Table 4.2. *Company footprint elements per layer*

Resource Layer	*Large automotive supplier*	*Automation equipment SME*	*Pump maker*
Socio Informal	Teams Entity Incentives Principles of Conduct	Production Circles Decentralised units with quick decision (uneven number), team work Incentive salary prop. PZG = (m*Te + Tr) *100/ At with m: Parts per period Te: time per part Tr: time per set-up At: time of presence	Flat hierarchy Decentralised units with quick decision mode (uneven number), team work Incentive salary, dependant on objective fulfilment Visualisation of objective fulfilment
Financial	Value stream Design	Standard cost per part	Entities Budgets
Informational	Visualisation of Progress	EDP supported JIT linked connections VBS (Electronic Kanban)	Kanban, EDP supported Kanban connections to supplier
Process	O.E.E:	Standard processes/parts: 6 hour, 12 hour, 48 hour product	Channel structure

configurations as well as on the activation of process sequences, where standards may provide for open-loop controls. For efficient adaptations as well as for continuous improvements of manufacturing networks, important control loops should be closed. The instruments suggested, the SoA and the criticality framework as outlined may be seen as possible next steps on the way to the closed-loop control of Dispersed Manufacturing Networks.

4.4.3 Extension and Breakdown of the Criticality Framework

Any manufacturing network, envisioned as consisting of self-similar fractals, carries the self-similarity properties for the attached models as well, as for the criticality framework with the embedded objectives' and resources' structures (Kühnle, 1995).

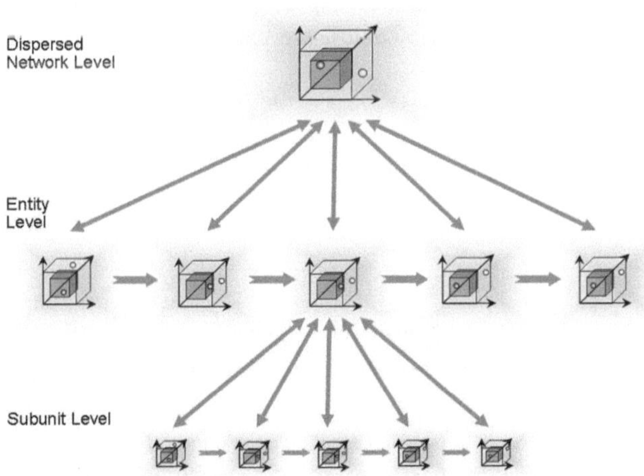

Figure 4.10. *Breakdown of Dispersed Manufacturing Network SoAs*

This means that the criticality framework in general and the dispersed networks' SoAs in particular may also be self-similarly broken down into entities and their respective levels of detail (Figure 4.10). This means that criticalities are not only immediately noticed but may also be more deeply analysed instantly.

Any critical state of an entity on a lower level may have an impact on the criticality of the involved unit as well as on the configuration of the total network. The ability to do quick, precise and reliable parameter settings, concerning the objectives as well as the resources' states is essential for any network control. Necessary reconfigurations, realignments or restructuring actions have to be control routine actions without significant reaction times. The criticality framework, as outlined, offers all these options and makes the management of Dispersed Manufacturing Networks much easier and more transparent, which can be seen in pilot cases of a multinational consumer goods supply network. Supported by a coordinator, executor, controller cycle that is activating in cases of criticality, all deviations are communicated to the other networks' units.

Arising criticalities are negotiated and harmonised with other units' objectives and resources as well as the total networks objectives in order to eventually obtain a consistent set of objectives as well as criticality thresholds (Figure 4.11). By adapting this procedure, the Dispersed Manufacturing Network is gradually optimised by continued negotiations and communication between units. The splitting or the removal of units from the network are decisions to be taken, if the criticality for an entity has to be repeatedly stated, i.e. the deviation cannot be eliminated by the unit's own efforts (self-optimisation).

Higher levels of detail are represented by SoAs "carrying" all SoAs of the lower network levels as (self-similar) folded structures. If there is an overall criticality for the unit (level u), the procedure has to be lifted up to the network level (level u+1), where the SoA of the total network, representing the aggregated criticality threshold, is addressed. Such adaptation cycles will be continued until all criticalities are removed and the (perhaps modified) objectives are achieved covering the highest level of the network. For these adaptations, especially the supports of the (self-similar) embedded structures on all levels of detail are extremely helpful. The objectives'

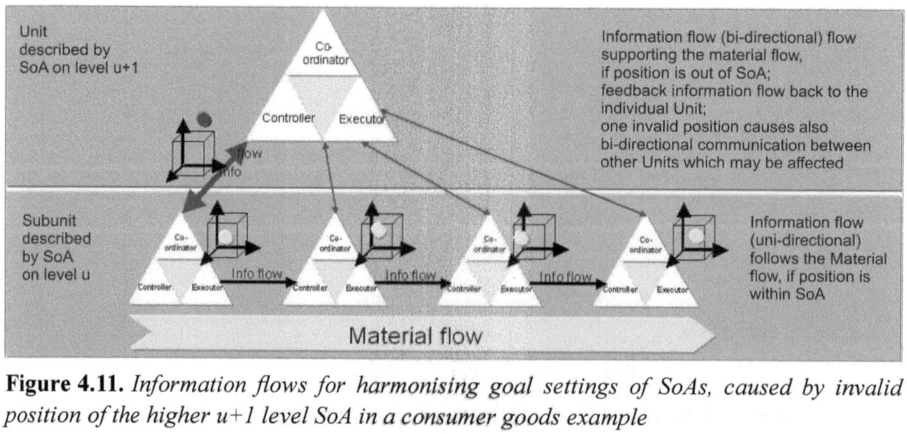

Figure 4.11. *Information flows for harmonising goal settings of SoAs, caused by invalid position of the higher u+1 level SoA in a consumer goods example*

and resources' bundles, represented by the SoA axes, allow further specifications of priorities in the objectives as well as the budgeting and the manipulation of the resources to be put in. For the criticality framework to be up to date at any time for all steps, the states and availabilities of the resources on all levels of detail must be quickly evaluated or budgeted for newly emerging structures of the network.

4.5 Conclusions and outlook

In the search for competitive excellence in production, Dispersed Manufacturing has received much attention during the last years, since understanding network characteristics in production obviously brings competitive advantages in everyday operations. Instead of trying to ignore or even eliminate structural behaviours of networks, Dispersed Manufacturing may successfully exploit these network properties by establishing adequate management procedures. Here, the concept of criticality, in synthesis with the self-similarity principle, proves to be a valuable contribution.

The criticality framework, as proposed, may also supply core procedures for the next generation production control software programmes supporting the control of Dispersed Manufacturing Networks. The given examples demonstrate the potential behind this concept with its strong support for communication and interlinking between units in Dispersed Manufacturing. In order to cope with frequencies and the unpredictability of criticalities, decision situations in Dispersed Manufacturing may be considerably simplified by introducing the self-similarity principle including all resulting description patterns. These simplifications have already been introduced for the generation control decisions and evaluations, e.g. on the task levels by software agent technology. At this level, the framework has widely proven to be extremely useful in Distributed Automation. The break down of Spaces of Activity and the embedded structures applies to all technical operations in manufacturing processes (McFarlane and Bussmann, 2000; Dietrich and Sauter, 2000; Kuehnle *et al.*, 2001). One generic concept for distributed manufacturing and order control, based on the described criticality framework, is PABADIS (Plant Automation BAsed on DIstributed Systems) (Lüder and Messerschmidt, 2005). PABADIS is serving as platform for leading manufacturing companies to define a new Open Industry Standard for distributed plant automation (AutomationML, 2007).

Comparable efforts are being made for further ERP developments. As the Distributed ICT world is about to be established on company levels as well, currently important vendors of ERP systems are intensively working on suitable solution packages, also making use of criticality and self-similarity principle, as outlined above. Contrary to some hard criticisms of managers, complexity approaches have indeed successfully generated important additional insights, which are novel and instructive. Many thoughtful managers of complex organisations will appreciate them as an enrichment of their production world-view in turbulent times. As in Dispersed Manufacturing, control and decision-making decentralise, and the "collaborative character" of the network is emphasised, the management becomes more complex. By exploiting the principle of self-similarity, this management complexity can be reduced because the entities of a network are able to respond better to increasingly volatile and competitive environments.

References

AutomationML (2007), available at: http://www.automationml.org/.

Barbasi, A-L. (2005), Network Theory: the Emergence of the Creative Enterprise, Science, Vol. 308, No. 5722, pp. 639-641.

Barlett, C.A., Goshal, S. (2000), Transnational Management, 3rd ed., McGraw-Hill, New York.

Bennett, D., Dekkers, R. (2005), Industrial Networks of the Future: A Critical Commentary on Research and Practice, in: Proceedings of the 12th International EurOMA Conference on Operational and Global Competitiveness, 19-22 June, Budapest, pp. 677–686.

Bratoukhine, A., Sauter, T. Peschke, J., Lüder, A. Klostermeyer, A. (2003), Distributed Automation: PABADIS vs. HMS. 1st IEEE Conference on Industrial Informatics, INDIN 03, Banff, Canada, September, Proceedings, pp. 294–300.

Budenske, J., Ahamad, A., Chartier, E. (1998), Agent-Based Architecture for Exchanging Modeling Data between Applications, Working Notes of the Agent-Based Manufacturing Workshop, Minneapolis, 11-14 Oct, pp. 2907–2912.

Camarinha-Matos, L.M., Afsarmanesh, H. (2005), Collaborative Networks: a new scientific discipline. Journal of Intelligent Manufacturing, Vol. 16, No. 4-5, pp. 439–452.

Citrin, J.M. , Neff, T.J. (2000), Digital Leadership, Managing, Vol. 18, pp. 42–50.

Clarke, C. (2005), Automotive Production Systems and Standardisation: From Ford to the Case of Mercedes-Benz, Springer, Berlin.

Cooper, R., Evans, S., Gregory, S., O'Brien, C. (2006), Design-Make-Serve, Delivering Products and Services to the Modern World, Discussion Document for the UK Manufacturing Professors Forum, July, London.

Davis, E. and Spekman, R. (2004), The Extended Enterprise: Gaining Competitive advantage through collaborative supply chains, FT Prentice Hall, New York.

Deen, S.M. (2003), Agent Based Manufacturing - Advances in the Holonic Approach. Advanced Information Processing, Springer, Berlin.

Dekkers, R. (2005), (R)Evolution, Organizations and the Dynamics of the Environment, Springer Science and Business Media, New York.

Dekkers, R., Luttervelt, C.A. van (2006), Industrial networks: Capturing changeability? Vol. 3, No. 1, pp. 1–24.

Dietrich, D., Sauter, T. (2000), Evolution Potentials for Fieldbus Systems, IEEE Workshop on Factory Communication Systems, Porto, pp. 343–350.

Dooley, K., Van de Ven, A. (1999), Explaining complex organizational dynamics, Organization Science, Vol. 10, No. 3, pp. 358–372.

e-Volution II (2004), Road map for e-business implementation in Extended Enterprises G1RD-CT-2002-00698, available at: http://www.e-vol.net/public/index.cfm.

GERAM IFIP/IFAC Task Force (1997), Generalised Enterprise Reference Architecture and Methodology, Version 1.5, 1997-09-27.

Gunasekaran, A., Yusuf, Y.Y. (2002), Agile Manufacturing, A Taxonomy of Strategic and Technological Imperatives, International Journal of Production Research, Vol. 40, No. 6, pp. 1357–1385.

Ivanov, D., Kaschel, J., Sokolov, B., Arldiipov, A., 2006, in IFIP International Federation for Information Processing, Volume 224, Network-Centric Collaboration and Supporting

Fireworks, eds. Camarinlia-Matos, L., Afsarmanesh, H., Ollus, M., Springer, Boston, pp. 15–22.

Jennings, N., Wooldridge, M. (1998), Agent Technology - Foundations, Applications, and Markets, Springer, Berlin.

Karasavvas, K., Burger, A., Baldock, R., (2005), A Criticality-Based Framework for Task Composition in Multi-Agent Bioinformatics Integration Systems, Bioinformatics, Vol. 21, No. 14, pp. 3155–3163.

Kauffman, S. (1993), The Origins of Order: Self-Organisation and Selection in Evolution, Oxford University Press, Oxford.

Klir, G.J. (1969), An Approach to General Systems Theory, New York .

Klostermeyer, A. (2002), Agentengestützte Navigation wandlungsfähiger Produktionssysteme, PhD Thesis, University of Magdeburg, Magdeburg.

Kosanke, K,, Zell, M. (2005), Interoperability, the INTEROP Project and Standardisation, in: Proceedings of the 11th International Conference on Concurrent Enterprising: Integrated Engineering of Products, Services and Organisations, Nottingham, pp. 49–51.

Kuehnle, H. (2007), A system of models contribution to production network (PN) theory, Journal of Intelligent Manufacturing, 18, Vol. 18, No. 5, pp. 543–551.

Kühnle, H. (2005), Fractal Extended Enterprise: Framework and Examples for multi-party supply chains, in: Shuping, Y., Xiaohui, C., Yu, Y. Modern industrial engineering and innovation in enterprise management. 12th International Conference on Industrial Engineering and Engineering Management, Chongqing, China, Vol. 1, pp. 211–217.

Kühnle, H. (1995), L'entreprise fractale. In: Braesch, C., Haurat, A. (Eds.), La modélisation systémique en entreprise, Pôle Productique Rhône-Alpes, Paris, pp. 263–272.

Kühnle, H., Klostermeyer, A., Lorentz, K. (2001), A Paradigm Shift to Distributed Systems in Plant Automation, in: Fares Sebaaly, M. Proceedings of The International NAISO. Congress on Information Science Innovations, March 17 – 2, Vol. 1, pp. 463–469.

Kuehnle, H., Wagenhaus, G. (2004), Extended enterprise architectures (E2A): towards a powerful mode of production, in: Gregory, M., Shi, Y. (Eds.): Designing and Operating Global Manufacturing & Supply Networks, 9th Annual Cambridge International Manufacturing Symposium, Cambridge, 20–21 September, pp. 168–183.

Lüder, A., Messerschmidt, R. (2005), Agent based Distributed Data Management for Factory Control, in: Proceedings of the 11th International Conference on Concurrent Enterprising - Integrated Engineering of Products, Services and Organisations, Nottingham, pp. 443–452.

Mandelbrot, B.B. (1982), The Fractal Geometry Of Nature, W.H. Freeman, San Francisco.

McFarlane, D.C., Bussmann, S. (2000), Developments in Holonic Production Planning and Control. International Journal of Production Planning and Control, Vol. 11, No. 6, pp. 522–536.

McKelvey, B. (2004), Simple rules' for improving corporate IQ: basic lessons from complexity science, in: Abdriani, P., Passiante, G. (Eds.) Complexity Theory and the Management of Networks. University of Lecce, Imperial College Press, Italy, pp. 39–52.

Merelli, E., Armano, G., Cannata, N., Corradini, F, d'Inverno, .M., Doms, A., Lord, P., Martin, A., Milanesi, L., Möller, S., Schroeder, M., Luck, M. (2007), Agents in bioinformatics, computational and systems biology, Briefings in Bioinformatics, Vol. 8, No. 1, pp. 45–59.

Noori, H., Lee, W.B. (2006), Dispersed network manufacturing: Adapting SMEs to compete on the global scale, Journal of Manufacturing Technology Management, Vol. 17, No. 8, pp. 1022–1041.

Olson, E.E., Eoyang, G.H., Beckhard, R., Vaill, P. (2001), Facilitating Organization Change: Lessons from Complexity Science, Jossey-Basss/Pfeiffer, San Francisco

Peng Y., Finin T., Labrou Y., Chu B. Long J. Tolone W., Boughannam A. (1998), A Multi-Agent System for Enterprise Integration, in: Proceedings PAAM, London, pp. 533–548.

Senge, P. (1990), The Fifth Discipline: The Art and Practice of the Learning Organization, Doubleday, New York.

Sporns, O. (2006), Small-world connectivity, motif composition, and complexity of fractal neuronal connections, BioSystems, Vol. 85, pp. 55–64.

Srai, J.S., Gregory, M.J. (2005), Supply Chain Capability Assessment of Global Operations, in: Proceedings of the 12th International EurOMA Conference on Operational and Global Competitiveness, 19–22 June, Budapest.

Teece, D.J., Pisano, G., Shuen, A. (1997), Dynamic Capabilities and Strategic Management, Strategic Management Journal, Vol. 18, No. 7, pp. 509–533.

van Brussels, H., Wyns J., Valckenaers P., Bongaerts L., Peeters P. (1998), Reference architecture for holonic manufacturing systems PROSA, Computers in Industry, Vol. 37, No. 3, pp. 255–274.

Warnecke, H.-J. (1997), The Fractal Company: A Revolution in Corporate Culture, Springer, Berlin.

Webb, C., Wohlfahrt, L., Wunram, M. Ziv, A, (2004), The Secrets of the Six Principles: A Guide to Robust Development of Organisations, Innovation Ecology, Pardesiya.

5 Collaborations in Industrial Networks: The Co-Evolutionary Perspective

Rob Dekkers
University of the West of Scotland

Abstract Currently, collaborations in production industry are experiencing a failure rate of 50%, if not more, and control of outsourcing is proving difficult; both findings from earlier research express the difficulties for managing networked enterprises. The lack of a problem-oriented understanding of the required systems setup and underlying control mechanisms might cause this. So far, academic research in management science has expanded models based on the individual company as an entity to address challenges posed by networks. Complementary approaches are required to match the specific characteristics of industrial and enterprise networks.

The application of principles of complex systems from natural sciences to collaborative enterprise networks as socio-technical systems might yield these complementary approaches. Five themes emerge from this point of view: dynamic description, coordination possibilities, integrative innovation, path dependency and information sharing. Interdisciplinary research should expand the available knowledge on the underlying mechanisms of collaborations by adopting models from the natural sciences (science of complex systems, network sciences, science of complexity, evolutionary models), and may possibly offer new perspectives to avoid traditional pitfalls (culture, leadership, trust) and to address the five themes.

Keywords Autopoiesis, Complex systems, Conceptual methods, Fitness landscapes, Interdisciplinary, Operations strategy, Systems analysis

5.1 Introduction

Engaging in collaborations and maintaining relationships within networks has become a major issue for industrial companies (Limerick *et al.*, 2002, p. 88). The possibilities of information technology and data-communication, the globalisation of markets and the ongoing specialisation of firms require adaptations by companies in the way they manage their operations. Particularly, these changes foster the specific characteristics of (international) networks of companies: collaboration to deliver products and services, decentralisation of decision-making amongst the agents and inter-organisational integration across companies involved to meet imposed performance requirements in competitive markets (adapted from O'Neill and Sackett [1994, p. 42]). That implies that management has moved away from the simple efficiency paradigm and related control processes – introduced in the 18th century by Adam Smith (1776) and by Frederick the Great of Prussia (Morgan, 1997, pp.

Table 5.1. *Evolution of organizational forms (Miles and Snow, 1984, p. 19). The table indicates the evolution of organization forms that are both internally and externally consistent. Miles and Snow state in their paper that a minimal fit is necessary for survival, that tight fit associates with corporate excellence, and that early fit provides a competitive advantage. Therefore, dynamic networks (industrial networks) require both internal fits and external fits, giving early adopters a competitive advantage.*

Period	Product-market strategy	Structure	Inventor or early user	Core activity and control mechanisms
1800 -	Single product or service. Local/ regional markets.	Agency	Numerous small owner-managed firms.	Personal direction and control.
1850 -	Limited, standardized product or service line. Regional/ national markets.	Functional	Carnegie Steel.	Central plan and budgets.
1900 -	Diversified, changing product or service line. National/ international markets.	Divisional	General Motors, Sears, Roebuck, Hewlett-Packard.	Corporate policies and division profit centres.
1950 -	Standard and innovative products or services. Stable and changing markets.	Matrix	Several aerospace and electronic firms.	Temporary teams and lateral resource allocation devices such as internal markets, joint planning systems, etc.
2000 -	Product or service design. Global, changing markets.	Dynamic network	International/ construction firms. Global consumer goods companies. Selected electronic and computer firms (e.g. IBM).	Broker-assembled temporary structures with shared information systems as basis for trust and co-ordination.

15-16) – to the notion of embedding companies in the environment and to the need for managing networks of agents. The overview by Miles and Snow (1984), see Table 5.1, indicates this move from simple paradigms to the more complicated forms of network-based organisations; even though published before the final transition could be witnessed. Especially, management in highly dynamic environments needs to address the resulting increasing complexity of industrial networked structures, contend Colotla *et al.* (2003, p. 1184). Hence, the shift to networked structures comes with new challenges for management and control.

Consequently, the complexity of industrial networked structures has attracted academic attention. After more than a decade of intensified research in the field of enterprise networks, the reasons for an increasing amount of networked enterprises have been broadly elaborated (e.g. Karlsson, 2003; Noori and Lee, 2002; Stock *et al.*, 1999; Zhang, 2000); most of these works emphasise emerging market opportunities. The need for capturing market opportunities has increased the significance of three main criteria of competitiveness: response time to customers, flexibility and innovation. Especially, the dominance of the response time (of both product development and engineering and the supply chain) and flexibility (product range and response to changes in demands) affects the operations of industrial networks. These characteristics have been associated with agility; in that respect, Goldman and Nagel (1993, p. 19) identified the twin characteristics of flexibility and response time as key contributors to agility while viewing companies as a network (note that they labelled response time with speed). Within industrial networks, the decreased response time might be mostly associated with the reduced lead-time for product development to capture product-market opportunities rather than solely for manufacturing and supply chain operations. Being able to seize those opportunities depends also on the capability to address customer requirements within industrial networks; a challenging area for both practitioners and researchers.

5.1.1 Background

A workshop in July 2002, attended by representatives from academic institutes and industrial companies, sponsored by the Dutch Ministry of Economic Affairs, confirms this stance; according to the outcomes of this workshop, the main issues for companies in relation to industrial networks focus on:

- Early supplier involvement.
- Value Added Networks.
- Optimisation across the total supply chain.
- Reduced time-to-volume during product development.
- Transparency of processes (primary and control processes).
- Customer-centred production (customisation).

In addition, a study by the National Research Council (Bollinger, 1998, pp. 3–4) indicates six grand challenges of which three directly refer to industrial networks: concurrent manufacturing, integration of human and technical resources, and reconfiguration of enterprises. All these issues and challenges refer both to the optimisation of the primary processes of the supply chain and to the response time of innovation and product development involving all agents as part of the network. The challenges and issues also indicate the lack of underpinning theory (similarly to Camarinha-Matos and Afsarmaneshi [2005] and Stadtler [2005]). Then, effectively addressing these issues and challenges depends on the involvement of all kinds of disciplines within management science, like technology management, outsourcing, partnering, supply chain management and organisational design. Past efforts by researchers into the domain of manufacturing organisations, which have extended networks of suppliers, have already yielded a wide range of concepts for decision-making on outsourcing (see Beaumont and Khan [2005] for an overview), Supply Chain Management (see Croom *et al.* [2000], the editorial overview by Gunasekaren *et al.* [2000], Stadtler [2005]), and technology management (or transfer). Alternatively, networked structures are increasingly based on the application of information and data-communication technologies, but have not been based on theoretical underpinnings, e.g. for Distributed Manufacturing (Dekkers, 2007). In addition, these concepts are mostly directed at the monolithic company and have been predominantly developed within existing disciplines of management science (Stadtler, 2005, p. 586). Hence, the direct application of theories and practices from these disciplines might not befit the characteristics of industrial networks. Therefore, the question emerges whether management science (and information technology) did gain ample insight and develop grounded theory to address the issues and challenges of industrial networked structures.

Although management science tries to describe the issues and challenges of industrial networked structures, no (new) theoretical base has been used yet. Firstly, direct transferences of these approaches in management science so far to networked enterprises regularly fail as they lack problem-oriented interdisciplinary inferences (for instance, driven by consilience [Wilson, 1998, pp. 8, 68]). The interdisciplinary approach has been identified as a fertile path to new insight. Secondly, the search for appropriate underpinning theory should align with the remark of Nassimbeni (1998, p. 539) that the bulk of available works is devoted to the contractual aspects and the social dynamics of inter-organisational relationships; he also claims that the dynamic forms of communication and coordination have been neglected. It is most likely, that attention for contractual and social aspects originates in the direct conversion from the hierarchical firm – with the direct control of resources and the belonging strategy towards suppliers – to the networks with more loosely connected entities; Camarinha-Matos and Afsarmaneshi (2005, p. 443) provide a similar argument. Thirdly, management science often takes the singular company as unit of analysis for identifying contingencies rather than searching for universal laws, as is usual in natural sciences. Putting it all together, the lack of interdisciplinary research focusing on dynamic forms of communication and coordination might explain the absence of underpinning theory for industrial networked structures; but might natural science help out through the mechanisms behind consilience?

In comparison to natural sciences and its research into dynamic phenomena, management science is a rather young discipline. Modern management sciences in Europe date back to the late 1970s, some of these were inspired by management cybernetics as theoretical base (e.g. Beer, 1972; Checkland, 1981). The approach of management cybernetics accepts the complexity of socio-economical systems and it searches for general principles for integrated systems control (Schwaninger, 2001). Object of research was the single, monolithic enterprise exposed to a stable environment, resulting in equilibrium-based theory. While the business environment has changed, management science still focuses on relatively static approaches. The science of cybernetic management has not fully accounted for the developments in the science of complexity. In this area of natural sciences a lot of progress has been made during the past decades. The theories in the science of complexity have been applied to the domain of management science (e.g. Houchin and MacLean, 2005; Stacey 1996), though some have criticised it (e.g. Burnes 2005; Romme and Witteloostuijn, 1997). At the same time, it might be a valuable contribution to understanding the dynamic phenomena in industrial networked structures, but that would require relating the science of complexity to the characteristics of these structures.

While the need for adequately addressing networked structures enfolded and subsequently for new approaches, management science has created theories that might have been adequate to some enterprises to deal with the contemporary challenges facing industry but not to others (Fischer and Hafen, 1997; Micklethwait and Wooldridge, 1996). The move to outsourcing as a consequence of core competencies might serve as an example in this respect; some authors responded by calling on insourcing (e.g. Lacity and Hirschheim, 1995). All these theories seem to have in common that their foundations stem from a variety of presuppositions pertaining to different factors that might directly influence the rate of success of an organisation at one place and time. Hence, instances or examples are taken as generalisation and normatively valid for a wider range of companies, as is the case for industrial networks.

5.1.2 Outline of Chapter

This chapter aims at presenting a new perspective to advance research in industrial networks by incorporating insight from sciences of complex systems (embedded in natural sciences). This contribution will explore to what extent concepts of natural sciences might aid to understand collaboration in networks. That advances in natural sciences will further our understanding of organisations has not gone unnoticed; sometimes these contributions have received appropriate attention because of the theory-model-based approach (McKelvey, 1999, p. 24) or they rephrased what was already known (Romme and van Witteloostuijn, 1997, pp. 68–70). Henceforth, this chapter builds on the exploration of concepts from natural sciences for management science and the identification of concepts that have value for industrial networks. Yet, it intends to outline how to arrive at new theory and a scientific base to address the challenges of industrial networks building on earlier findings in case studies on outsourcing and collaborations. It seems that we are approaching the limits of the more static approaches directed at the singular company; hence, a new approach should be generated to address the **dynamics** of the business environment and the complexity of industrial networks (Bouvier-Patron, 2001, p. 22). In its time, cybernetic management boosted management science. Its successors in the natural sciences, e.g. the science of complexity and network science, might shed light on the dynamic phenomena related to industrial networks: a promising shift in underlying paradigms?

Thereto, this chapter will briefly outline how a change in the theoretical base for industrial networks might lead to more predictive models to describe typical phenomena. Firstly, the Section 5.2 will focus on the validity of existing models for the domain of networks based on findings in the domain of outsourcing and collaborations; this results in challenges for the research. Then the chapter explores the nature of industrial networks as socio-technical systems (Section 5.3). Furthermore, it will discuss the advances in three strands of natural sciences (network sciences, science of complexity and co-evolutionary systems) and their meaning for improving insight in industrial networks. Finally, the chapter will present outlines of challenges to develop new models for processes and structures in industrial networks based on dynamic forms of communication and coordination.

5.2 Outsourcing and Collaborations as Source for Increased Complexity

First, this section will focus on the existing models for the domain of industrial networks, characterised by the static perspective. This perspective – typical for many areas in management science – is found especially in approaches and methods for outsourcing, partnerships, alliances as a make-or-buy decision (e.g. Cáñez *et al.*, 2000; Humphreys *et al.*, 2002; Probert, 1997) or a strategic decision-making related to competencies (Dekkers, 2000; Hafeez *et al.*, 2002). Generally, this perspective leads to one-time decision-making and ignores the continuously acting dynamics of the environment. The reduction of transaction costs (originating in transaction cost analysis [Williamson, 1975]), enabled through technological improvements in communications and logistics during the last years, entails an increasing outsourcing tendency in production industry while increasing inter-firm complexity for business process coordination.

5.2.1 Control of Outsourcing

How does industry cope with the increasing scope of outsourcing? An exploratory investigation into practices in six companies points to poor control of outsourcing (Dekkers, 2005a). Most of the six case studies – companies with primary processes mostly based on Engineering-to-Order and Make-to-Order - experienced problems with implementing manufacturing strategies and meeting performance requirements. Firstly, part of the manufacturing strategy of the companies should have addressed the core competencies and the opportunities for outsourcing. All companies, except one, had done so, implicitly or explicitly. Mostly this strategy had not been transferred to guidelines for implementation, which is why that decision-making on outsourcing occurred randomly or by opportunity. In addition, there was no feedback about supplier performance to the stages of design and engineering, therefore sometimes problems would recur over and over again. Furthermore, none of the companies reported following an active approach towards the supplier networks to expand the technological capabilities. On top of that, the operational control of outsourced activities posed additional challenges, although not all companies were aware of the impact. All companies reported problems on in-time deliveries by suppliers; some of the problems arose from reactive interventions rather than pro-active securing of purchasing orders. In two cases, the in-house production of some manufacturing processes proved more beneficial than outsourcing on hindsight. Putting it all together, the operational control posed a wide variety of problems showing inadequate control mechanisms and a poor integration between design, engineering, purchasing and manufacturing.

These *over-the-wall tactics of complexity* through outsourcing do not lead to a reduction of structural complexity within the monolithic enterprise. Scientific Management, inspired by Taylor at the beginning of the 20th century, pinpoints the underlying effect: the complexity of a task decreases by dividing it into several sub-tasks while increasing coordination efforts. Assuming that these sub-tasks need specific competencies, they are allocated to different partners within a networked enterprise. At the same time, a significant increase is needed in allocated resources for coordination and control of product development and manufacturing, calling for adequate **coordination mechanisms**.

5.2.2 Collaborations in Manufacturing

Consequently, studies reveal that at least 50% of all collaborative projects in manufacturing industry fail (e.g. Bierach 2000; Jansen and Petersen, 2000; Wognum and Faber, 2001, p. 32). Multiple reasons might attribute to this high rate of failure:

- the often postulated heterogeneity in networks has lead to a lack of guidance and control, rendering these networks participative but dissipative (supported by Huggins [2000, p. 132]). Hierarchical structures could be helpful but are not part of the postulated network paradigm;
- the ignorance about what type of network architecture and what type of control fits to what type of specific collaboration. Most initiatives in network architecture are driven by information technologies (e.g. Maturana and Norrie, 1996; Ryu and Jung, 2003) but these do not necessarily address issues of collaboration (see Dekkers [2007] for the case of Distributed Manufacturing). Lamming *et al.* (2000) propose a classification based on the complexity and the degree of innovativeness of products but that only leads to abstract descriptions of networks (insufficient for the design of a network architecture);
- the lack of knowledge about what accounts for a networks' complexity. In that sense, Lamming *et al.*'s classification provides some insight. Models and instruments for managers to handle the internal complexity of networks and the imposed complexity by the environment are not available, yet;
- the unforeseen, emergent network effects in elasticity, controllability and overall network and production system behaviour.

Even with these detrimental effects, the benefits of the pending boom of division of tasks among network members still promises to be greater than the challenges it causes; especially, increasing the rate of innovation and improving performance through shorter response-times and increased flexibility.

This is demonstrated through the increasing participation of SMEs in industrial networks and virtual organisations (Bennet and Dekkers, 2005)[1]. It might well be that not only advances in information and data-communication technology facilitate this transition but also the capability of these networks to address response time and flexibility, key attributes of agility. The application of the function-oriented concept *Virtual Factory* (Schuh, Millarg and Görnasson, 1998) might be considered

1 *See also Chapter 2.*

one of the first operational concepts for networked SMEs by centering on the way cooperation is built as a socio-technical system. The concept provides guidelines for structuring the cooperation process so that ad hoc cooperation can be built up quickly and flexibly. The concept contains two different layers: the stable platform and the dynamic networks. Recent studies show that not all effects and interrelations have been integrated in the concept (Schuh and Wegehaupt, 2004a, b). A lack of management functions and existing trust has been identified as major pitfalls in the four analysed systems. Besides an inadequate competence pooling, the missing fit between the products being produced and the network's internal complexity of the order-processing leads to inefficiencies. The original concept has consequently been adapted. A business processor, executing tactical management functions like project management, mediation, capacity planning or product data management, has therefore been introduced as a socio-technical hybrid entity (Wegehaupt, 2004). The potential of this operational concept is not limited to production but can also be implemented to highly dynamic projects in product development (Schuh and Sauer, 2005a, b); hence addressing the performance requirements: response time, flexibility and innovation (and thus achieving agility).

5.3 Challenges for Research

The abridged problems of outsourcing and virtual organisations emerge due to a lack of knowledge in the flexibility, the adaptation and the dynamics of networked enterprises. It might well be that to date, network sciences and the science of complexity have not resolved the contemporary problems of production industry sufficiently; especially because they have not been connected to the three characteristics of industrial networks (collaboration to deliver products and services, decentralisation of decision-making amongst the agents and inter-organisational integration across companies involved to meet imposed performance requirements). Consequently, this chapter takes as point of departure that the theories of complex systems, accounting for these characteristics and found in natural sciences – biology, physics or chemistry – can still be considered the upcoming paradigm for the phenomena of networks. Whereas, with regard to management science, the complexity paradigm inversely constitutes a new theory for understanding enterprise networks and promises potential for major progress when dealing with socio-technical systems. The emerging challenge for management science resides in developing models and methodologies for holistic complexity management in new networked structures that come forward (much like the call for the theory-model link by McKelvey [1999, p. 24]).

5.3.1 Complexity in Socio-Technical Systems

But what constitutes complexity in a socio-technical system? According to the constructivist school of Watzlawick (1976) and von Foerster (1985), the human mind "constructs" a model of the environment to cope with the complexity of measurable parameters (much like the theory-model link that McKelvey [1999, pp. 16–18] introduces). This method is a valuable step towards the reduction of complexity and thus a question of efficiency in "reality management"; an approach similar to Soft Systems Methodology (Checkland, 1981) and the Viable Systems Model (Beer, 1959). According to Riedl (2000), the human senses and mind are archaic instruments for coping with reality from less complicated and complex times. He postulates that now with an ever increasingly complex environment the mere capacity of the apparatus for stimulus processing is no longer sufficient to resolve fully the complexity of a situation. Consequently, any reduction of complexity, i.e. simplification, aims at creating a manageable level for the humans bestowed with the task of controlling complex socio-technical systems. The degree of simplification strongly depends on the amount, interdependence and behaviour of inherent core complexity drivers that represent the underlying structure (static dimension of complexity) and behaviour (dynamic dimension of complexity) of the socio-technical system.

Alternatively, research into industrial networks might aim at increasing the Complexity Handling Capability of the individual organisations and the networks they participate in. No matter how companies build on existing capabilities, present in available resources and current structures, choices for coping with the external changes remain limited. Ashmos *et al.* (2000, p. 589) demonstrate that by examining eight cases and inferring that a complexity reduction response is inferior to a complexity absorption response. In the end, companies will have to adapt to the external changes through increasing the Complexity Handling Capability, which means building on existing capabilities for new situations or incorporating new knowledge for creating new capabilities (Boswijk, 1992, p. 100; Bouvier-Patron, 2001, pp. 25-28). Much of the complexity – i.e. the new challenges – is imposed by the environment on the

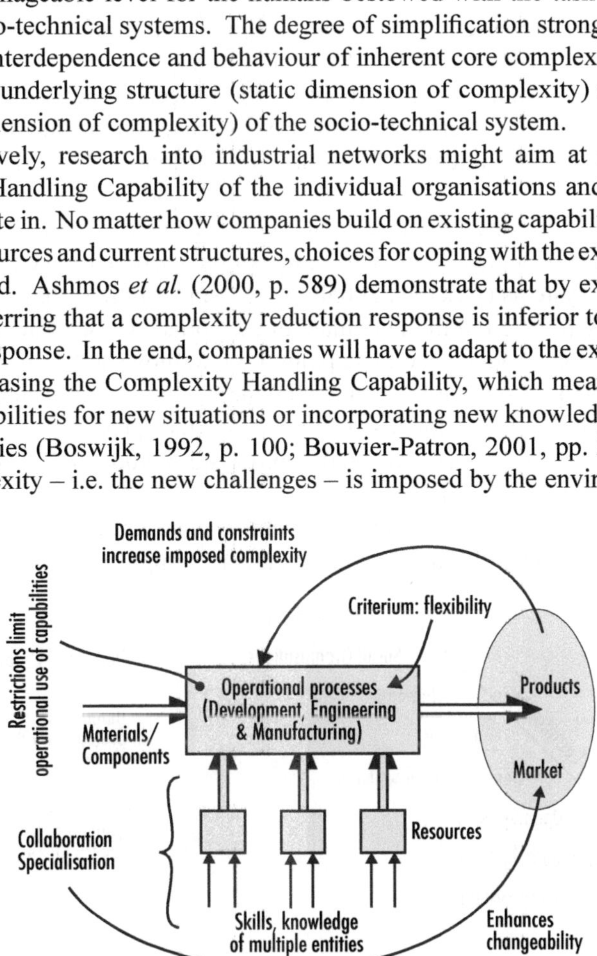

Figure 5.1. *Changes affecting industrial companies and networks (Dekkers and van Luttervelt, 2006, p. 6). Response time and reliability of delivery increase the complexity of internal processes. Secondly, the wider availability of resources claims more attention for organisational integration, interaction between organisational entities and adaptation by collaboration and specialisation. Together with the call for innovation and customisation the externally imposed complexity increases.*

networks and their elements, the individual organisations. Hence, the challenge for complexity research in enterprise networks is the identification of adequate forms of system representation, the analysis of interdependence among core elements, and the specification of complexity drivers, accounting for the dynamics and complexity imposed by the environment (Figure 5.1).

5.3.2 Networks as Socio-Technical Systems

That brings us to General Systems Theory, which copes with representing systems and characterises organisations as open, dynamic, purposeful, productive socio-technical systems (Ulrich, 2001). Several approaches within General Systems Theory exist, e.g. Maturana and Varela (1980), Beer (1966; 1972), describing generic organisational concepts. Ropohl (1999) focuses on the integration of social systems and technical systems within three dimensions. He distinguishes three subsystems: the action system, the execution system, and the goal setting system, as the dimensions of the inner structure of any socio-technical system. A methodology, known as the Delft School Approach, designs organisational structures viewed as socio-technical systems (Dekkers, 2005b, p. 7). None of the systems theories have been adequately transferred to the domain of networks. These systems theories, to model organisations from a cybernetic point of view, move at the third system level of Boulding (1956), see Figure 5.2, and combine it with a socio-technical approach for designing new organisational structures. Yet, the questions remains whether General Systems Theory has generated appropriate methodologies for describing and analysing networked structures.

Since organisations and networked structures represent the eighth system level in Boulding's hierarchy (1956), the systems theories might need elaboration by the

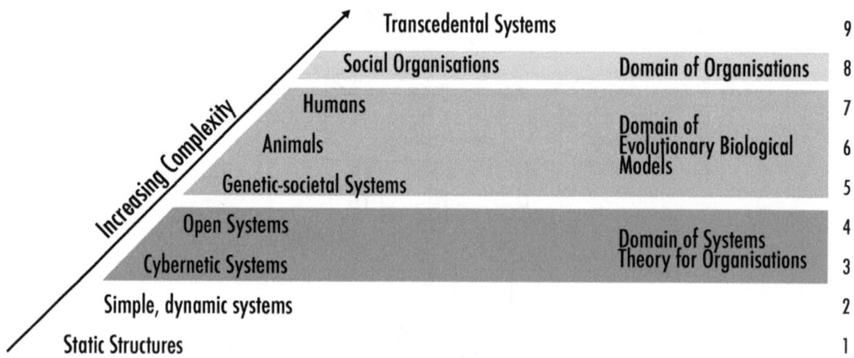

Figure 5.2. *The nine levels of Boulding (1956) (Dekkers, 2005b, p. 69). The domain of organisations moves at the eighth level indicating the importance of meaning, value systems and symbolisation. The domain of systems theory and some other approaches in management science (e.g. information technology) find themselves at the third and fourth levels. Models from complexity science, network science and evolutionary biology might bridge the gap between some of the approaches in management science and the actual organisational domain.*

adoption of theories of complex systems, networks and biological models. At the level of organisations, the validity of the design approach might be questioned, which is typical for the third and fourth levels. The design approach has the characteristics of static, one-time interventions, which companies might have to avoid because of their severe effects on organisations (Dekkers, 2005b, pp. 238–239). The review of other theories, like those of complex systems, networks and co-evolutionary models, might facilitate the search for the structures of organisations and industrial networks, needed for adapting to changes in the environment and continuous change (as demonstrated in Dekkers [2005b]). These theories move closer to the eighth level of Boulding and therefore they might constitute more adequate theories for organisations and industrial networks.

In addition, human-influenced complex networks, e.g. the World-Wide Web, human acquaintance networks, have common properties for connectivity, which are hardly compatible with existing cybernetic approaches at the third and fourth levels. The so-called *small-world property*, the most known of these specific properties, states that the average path length in the network is small relative to the system size (Milgram, 1967). This phenomenon was already scientifically studied more than three decades ago, long preceding its notoriety. In fact, the phrase *six degrees of separation* (Guare, 1990), another popular slogan depicting the small-world phenomenon, is due to Milgram's 1967 experiment. Another property of complex networks is clustering, i.e. the increased probability that pairs of nodes with a common neighbour are also connected. Since then, increased efforts have been dedicated to identify other measures of complex (enterprise) networks (Fricker, 1996). Perhaps, the most important is the *distribution of degrees*, i.e. the distribution of the number of links the nodes have. It has been shown that several real world networks have scale-free distributions, often in the form of a power law. In these networks, a huge number of nodes have only one or two neighbours, while a couple of them are massively connected. These three specific properties (small-world property, six degrees of separation, distribution of degrees) strongly influence the behaviour of agents in networks and the development of these networks.

The properties have been translated into mathematical models and applications focusing on large networks and connectivity (e.g. Klemm *et al.*, 2003; Krapivsky and Redner, 2001; Newman, 2003; Watts and Strogatz, 1998); most of these applications show that these properties make networks behave more dynamically. Industrial networks consist of a limited number of agents and therefore, might display other behaviour than large networks. Take, for example, the industry sector for flow-wrapping packaging equipment that only consists of 350 companies world-wide. The expansion to industrial networks should include the behaviour of agents (not just agents as nodes) and the development of traits for selection. While a number of models have been proposed to generate networks with some combination of the three properties above, virtually each of these describes a process that ends with a network having the desired properties. Less effort has been devoted to the design of a dynamic system that would generate and maintain such a network. While there exist few such models (e.g. Friedli, 2000; Schwaninger, 2000), most of them assume that the system size or the number of links increases. Therefore, advances in theories for industrial networks should focus on the dynamics of socio-technical systems accounting for the typical properties of complex networks.

5.3.3 Network Science

Latest empirical surveys on successful networks (Bonabeau and Meyer, 2001; Car *et al.*, 2002; Dannenmaier *et al.*, 2003; Kennedy *et al.*, 2001) give strong hold to the hypothesis that only two different paradigms of control in networks exist (see Figure 5.3):

- the first is the paradigm of guided networks which comprises features of hierarchical control in terms of first order cybernetics with the controller being a constituent element of the system. This type's fundamental specifications are hierarchical networks and focal networks. Guidance is realised in guided networks by explicit planning of interaction in advance to execution (also supported by Perona and Miragliotta [2004, p. 113]);
- the second paradigm is the self-organised, organic network which is implicitly managed by Adam Smith's invisible hand of an external context, not being explicitly a control element of the system (second order cybernetics); one might differentiate between dyadic and triadic networks. Inherent to its character of local intrinsic-triggered interaction, self-organised networks can only be managed in an implicit way, which means non-deterministic coordination of activities. Actively conditioning the network's context by establishing an effective rule setting canalises network-activities towards a specific aim. Synergistic effects emerge by making the network's entities acting in a congruent way holding up a necessary level of efficiency. By doing so, the network's entities adapt their own complexity (i.e. activities, structure, behaviour) to the external requirements as parameters of their context (Choi *et al.*, 2001, pp. 364–365). Thereby a global order emerges as a result of congruent local interactions (Stacey, 1993b).

	Guided Hierarchical Networks / Focal Networks	***Self-organized, emergent network*** Dyadic / Triadic
Characteristics	▪ Two levels: Network and Collaboration ▪ Explicitly managed (focal entity) ▪ Configuration and Planning of interaction in advance to execution/enactment ▪ Span of control > 1	▪ One level: Collaboration as Network ▪ Implicitly managed (context) ▪ Instantaneous, local dyadic interaction [self-organization ▪ Span of control = 1 (self-reference)
Field of Application	▪ Design and Product Development ▪ Production (Manufacturing, Assembly) Batch-Production / Make-to-Order, One of a Kind	▪ Innovation during ideation phase ▪ Procurement ▪ Sales & Marketing ▪ Service

Figure 5.3. *Control paradigms for networks. The hypothesis that only two fundamentally different types of networks do exist is displayed. Each type of network generates a different type of complexity and suits different contingencies.*

Depending on the type of problem one of these two paradigms is effective. All intermediary forms either fail or evolve to either of the two forms over time (Wegehaupt, 2004, p. 24). The more constructivist approach of guided networks especially suits product development and production (the search for concepts for **integrated innovation**); the implicit management's potential resides in aligning ideation activities, procurement and service. Creativity and effectiveness are more important than efficiency for innovation processes, which constitutes a paradigm for traditional technomorphous milestone concepts. Note that this distinction follows the classification into mechanistic and organistic organisations by Burns and Stalker (1961). Hence, stability and instability issues of industrial networks might be driven by factors related to appropriate network control, although these driving factors have not been established, yet.

The guidance paradigm is well established in cybernetic approaches for system control, systems engineering or management cybernetics. An example is the Delft School Approach, partially based on socio-technical inferences. The paradigm has found its most popular representation in the works of the biologist and philosopher Stafford Beer, whose insights in complexity management are based on Ashby's Law of Requisite Variety (Ashby, 1956) and culminated in the Viable System Model (Beer, 1972). This model is until today the only generic and concise representation of the architecture for structuring purposeful systems. Not only does it yield unprecedented insight into the field of complexity management, but it is also an unprecedented success of interdisciplinary research.

Taking into consideration the different dimensions of complexity in socio-technical systems, the two network and control paradigms entail different types and patterns of complexity for tackling different types of collaborative problems. Ashby's Law of Requisite Variety postulates that only complexity can absorb complexity (Ashby, 1956). Abiding this requires matching the collaborative system's variety (behaviour) with the complexity of the problem to be solved, henceforth increasing the Complexity Handling Capability.

5.3.4 Science of Complexity

With the proliferation of the network paradigm the hierarchical approach towards control has lost its charm and attention in science. Inspired by the Zeitgeist of the late 1980s, the trend of decentralization and the postulation of non-hierarchical, participative and distributed control in society and organisations also penetrated complexity science (Malik, 1992). Starting with the works of the Santa Fe Institute in the early 1980s, the paradigm of self-organisation emerged and opened a new branch in the explanation and control of complexity (Jost, 2004). With the increasing number of elements in artificial systems – turning them into net-like entities – their control became increasingly complex (Tucker *et al.*, 2003). This made the deterministic, top-down approach to systems control inefficient, if not impossible, especially against the background of a highly dynamic environment.

It is believed that even in the study of complexity simple and therefore comprehensible laws exist. The field of study of complex systems holds that the dynamics of complex systems are founded on universal principles that may be used to describe disparate problems ranging from particle physics to the economics of societies (Kauffman, 1993). The development of complexity science offers a shift in scientific approach with the potential to profoundly affect business, organisations and government. Complexity science strives to uncover the underlying principles and emergent behaviour of complex systems. Complex systems are composed of numerous, varied, simultaneously interacting agents. The goal of complexity science is to understand these complex systems – what "rules" govern their behaviour, how they adapt to change, learn efficiently and optimize their own behaviour.

The term complexity can be understood in two ways relevant to this research:

(a) as an expression of structure, mostly internally oriented either part of networks or an individual system;
(b) as an expression of emergence, more rooted in new behaviour and complexity imposed by the environment.

Internal complexity can be seen as a design parameter (Perona and Miragliotta, 2004. p. 104), even though not sufficiently defined in cybernetic approaches. To cope with emergence, different entities might develop different types of Complexity Handling Capability; under these conditions, balance will hardly be achieved, only paradigms that address the dynamics of industrial networks and the environment will elect for elaboration within the context of this research. In an organisational context, complexity provides an explanatory framework of how organisations behave, how individuals and organisations interact, relate and evolve within a larger social ecosystem. Complexity also explains why interventions may have unanticipated consequences (Buchanan, 2004). The intricate interrelationships of elements within a complex system give rise to multiple chains of dependencies.

The theory of complex adaptive systems as the state of the art of research in the field of self-organisation cannot be assigned to one particular field of science (Nieuwstad, 1997). However, it has found its way to many adjacent disciplines, e.g. evolutionary computation, evolutionary biology (e.g. Bonabeau, 1998), technology management. Self-organisation, as a general theory for complex systems, is considered a new paradigm and a fundamental challenge to the traditional, linear and deterministic programme in science as a whole and its ideas of certainty and randomness. In complex adaptive systems that involve large numbers of entities, emergent, global behaviours that arise from localized interactions are a critical concept. Understanding and shaping emergence may be essential to such systems' success; from this perspective, phenomena difficult to comprehend have found explanations that yielded more appropriate insights.

5.3.5 Industrial Networks as Co-Evolutionary Systems

Even within the domain of biological (evolutionary) models, a larger number of theories exist that might describe adequately the existence of industrial networks and collaboration, for example, ecosystems. In biology, co-evolution, as an adequate description for collaboration, is the mutual evolutionary influence between two species that become dependent on each other. These concepts from evolutionary

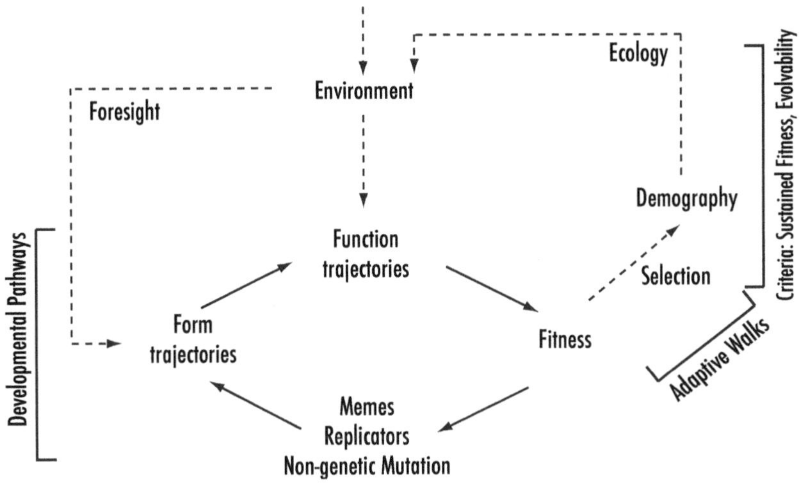

Figure 5.4. *Evolutionary mechanisms for organisations as reference model. Memes and replicators serve as input for genetic formation, which exists besides non-genetic formation. Developmental pathways determine the form and function trajectories. These pathways also relate to organisations being a class of allopoietic systems (structurally closed). The selectional processes select beneficial phenotypes on fitness following adaptive walks. Organisations have the capability of foresight in contrast to organisms.*

biology cover a wide range of interaction between agents, for example altruism. Within the domain of industrial networks, the mutual dependence has been recognised as a potential direction for collaboration. Wiendahl and Scholtissek (1994, p. 539) mention co-evolution as a possible avenue for research since it enables the comprehension of how agents evolve and adapt continuously in the context of networks. Given that this is true, how might collaboration evolve in industrial networks?

That this view might hold perspectives, other management scientists have discovered, like Lewin and Volberda (1999). They focus on the emergence of new organisational forms (Lewin *et al.*, 1999), without clearly defining the "organisational form" (McKendrick and Carroll, 2001, p. 662). Co-evolution has appeared in writings that build on the work of Nelson and Winter (1982). For the purpose of this chapter, it suffices to remark that these models do not address the intertwined cycles of the reference model in Figure 5.4. Especially, the absence of fitness landscapes is lacking, which limits the validity of the outcomes; hence, these models are so far rendered inappropriate for describing co-evolution in networks. Co-evolution, when used in its sense of the mutual development of organisms, benefits selectional forces (i.e. survival on the long-run). Volberda and Lewin (2003) touch on this but do not exploit the resemblance between organisms and organisations. Thus the description of co-evolution starts with fitness landscapes as an expression of the fitness of the belonging genotypes.

Fitness resembles height, a measure for expressing the fitness of a genotype, equal to thoughts of Wright's adaptive landscape (Wright, 1982)[2]. Fitter genotypes are higher than less fit genotypes. Consider a genotype with only four genes, each having two alleles: *1* and *0* (i.e. a Boolean representation of the state of each gene), resulting in 16 possible genotypes, each an unique combination of the different states of the four genes. Each vertex corresponds to one of the 16 possibilities (see Figure 5.5). Each vertex differs only by one mutation from the neighbouring ones, representing the step of a single mutation, thereby stating that each mutation as such is independent from the state of the other genes. Each genotype is arbitrarily assigned to a fitness value ranked from worst (i.e. *1*) to best (i.e. *16*), representing a peak in the fitness landscape. An adaptive walk might begin at any vertex and move to vertices that have higher fitness values. An adaptive walk ends at a local optimum, not necessarily the highest optimum, a vertex that has a higher fitness value than all its one-mutant neighbours. In the figure, it is shown that three local optima exist at which adaptive walks may end.

As Kauffman (1995, pp. 166–167) states, evolution requires landscapes that are not random. In fact, on random landscapes, finding the global peak by searching uphill is useless; we have to search the entire space of possibilities. From any initial point on a landscape, adaptive walks reach local peaks after some number of steps. Additionally, no matter, where an adaptive walks starts, if the population is allowed to walk only uphill, it can reach only an infinitesimal fraction of the local peaks. However, in reality, the fitness landscapes that underlie the mutation steps of gradualism are correlated, and often local peaks have similar heights.

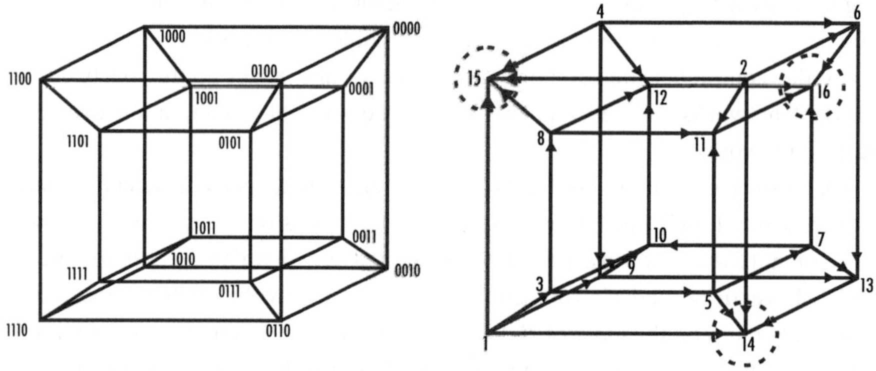

Figure 5.5. *The N-model as proposed by Kauffman (1993, p. 38). 16 possible peptides of length 4 aminos are arranged as vertices on a four-dimensional Boolean hypercube. Each peptide connects to its four one-mutant neighbours, accessible by changing a single amino acid from 1 to 0 or from 0 to 1. The hypercube on the left represents this four-dimensional peptide space. In the hypercube on the right-hand side, each peptide has been assigned, at random, a rank-order fitness, ranging from the worst, 1, to the best, 16. Directions of such moves between adjacent positions are shown by arrows from the less fit to the more fit. Peptides fitter than all one-mutant neighbours are local optima (three in this case).*

2 *Wright's writings about adaptive landscapes date back to 1932.*

1	2	3	w1	w2	w3	W
0	0	0	0.6	0.3	0.5	0.47
0	0	1	0.1	0.5	0.9	0.50
0	1	0	0.4	0.8	0.1	0.43
0	1	1	0.3	0.5	0.8	0.53
1	0	0	0.9	0.9	0.7	0.83
1	0	1	0.7	0.2	0.3	0.40
1	1	0	0.6	0.7	0.6	0.63
1	1	1	0.7	0.9	0.5	0.70

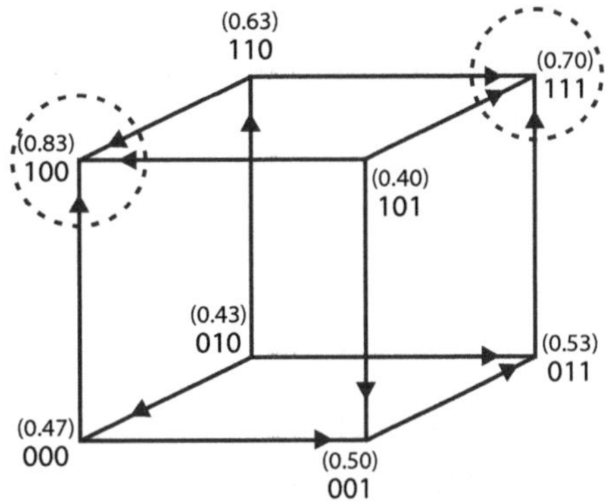

Figure 5.6. *NK-model as developed by Kauffman (1993, p. 42). In the upper left corner it shows the assignment of K=2 epistatic inputs to each site. The assignment of the fitness values to each of the three genes 1, 2 and 3. These fitness values then assign fitness to each of the 23=8 possible genotypes as the mean value of the fitness contributions of the three genes. The fitness landscape on the three-dimensional Boolean cube corresponding to the fitness values of the eight genotypes is depicted. More than one local optimum exists.*

Rugged landscapes are those landscapes in which the fitness of one gene depends on that part and upon *K* other parts among the *N* present in the landscape (Kauffman, 1993, p 40). This NK-model offers further insight in the mechanisms of evolution and selection (Kaufmann, 1993, pp. 40–54). Again consider an organism with *N* gene loci, each with two alleles, *1* and *0*. Let *K* stand for the average number of other loci, which epistatically affect the fitness contribution of each locus. The fitness contribution of the allele at the ith locus depends on itself (whether it is *1* or *0*) and on the other alleles, *1* or *0*, at *K* other loci, hence upon *K+1* alleles. The number of combinations of these alleles is just 2^{K+1}. Kauffman selects at random to each of the 2^{K+1} combinations a different fitness contribution from a uniform distribution between *0.0* and *1.0* (see Figure 5.6). The fitness of one entire genotype can be calculated as the average of all of the loci.

Despite the importance of fitness landscapes for evolutionary processes, Kauffman (1995, p. 161) states that biologists hardly know what such fitness landscapes look like or how successful a search process is as a function of landscape structure. The landscapes may vary from smooth, single-peaked to rugged, multi-peaked landscapes. During evolution, species search these landscapes using mutation, recombination, and selection, a process for which the NK-model provides insight into particular phenomena accompanying the adaptive walk.

The mechanisms of co-evolution will affect the adaptive walks. Kauffman (1993, pp. 243–245) extends the NK-model to co-evolution by adding the constraint that each trait in species *1* depends epistatically on *K* traits internally and on *C* traits in species *2* (the so-called NK[C]-model). More generally, in an ecosystem with *S* species, each trait in a species will depend on *K* traits internally and on *C* traits in each of the S_i among the *S* species with which it interacts. Therefore, if one species adapts, it both changes the fitness of other species and deforms their landscapes.

The coupling of the fitness landscapes will affect the search for increased fitness. The expansion of a network or more integration will affect the adaptive walks (Kauffman, 1993, pp. 252-253). When a new link is introduced (i.e. increasing *K*) the genetic locus spreads throughout a population in three ways: (a) the new epistatic link, when it forms, causes the genotype to be fitter, (b) the new epistatic link is near neutral and spreads through the population by random drift, or (c) the new link not only has a direct effect on the fitness of the current genotype but also increases the inclusive fitness of the individual and its genetic descendants. It suggests that optimisation in co-evolutionary dynamics becomes possible by optimisation mechanisms that search for optimal traits in relation to the coupled traits. The second option for a network consists of increasing its reach, which compares to increasing the number of species *S*. When that happens the waiting time to encounter equilibrium increases, the mean fitness of the co-evolving partners decreases, and the fluctuations in fitness of the co-evolving partners increase dramatically. The increase of agents might lead to a new optimisation in traits and coupled traits but only after going through a period of instability; in addition to instabilities, these phenomena point to **path dependency**: mutational steps in fitness landscapes, depend on previous steps and determine future routes for increasing fitness.

These instabilities might come with phase changes, or percolation, in the Boolean networks captured in the NK-model (Kauffman, 1995, pp. 80-92). Four particular states arise when the NK-model is analysed for the principles of self-organisation. Firstly, at $K=1$, the orderly regime appears, in which independent subsystems function as largely isolated islands with minimal interaction. At $K=2$, the network is at the edge of chaos, the ordered regime rules at maximum capacity but chaos is around the corner. At values ranging from $K=2$ to $K=5$ the transition to chaos appears although indications are that this transition happens already before $K=3$. From $K>5$, the network displays chaotic behaviour. All these four possibilities of *K* indicate that the behaviour of networks varies strongly according to the epistatic links between traits.

But not only the epistatic links determine co-evolution, symbiosis also could assist in understanding collaboration in networks. The concept of symbiosis deserves some more attention as a form of co-evolution in networks; symbiosis in this chapter is viewed as an extension of the NK[C]-model (*C* as parameter of interconnectivity between different agents). Symbiosis is an interaction between two different organisms living together in more or less intimate association or even the merging of two dissimilar organisms. The various forms of symbiosis include:

- parasitism, in which the association is disadvantageous or destructive to one of the organisms and beneficial to the other;
- mutualism, in which the association is advantageous to both;
- commensalism, in which one member of the association benefits while the other is not affected;
- amensalism, in which the association is disadvantageous to one member while the other is not affected.

In some cases, the term symbiosis is used only if the association is obligatory and benefits both organisms. Sometimes, altruistic behaviour benefits another organism, not necessarily closely related, while being apparently detrimental to the organism performing the behaviour (Trivers, 1971, p. 35). Symbiosis as defined in this chapter does not restrict the term to only the mutually beneficial interactions. It has strong similarities to the coupling of the traits in the NK[C]-model to describe co-evolution and therefore to complex systems (e.g. Sigmund, 1998). It indicates that the mutual relationships have at least two dimensions: the fitness of each of the two agents involved.

5.3.5 Autopoiesis

Similar to the mutual relationships of symbiosis, Khanna *et al.* (1998) have used the terms private and common benefits. They state that in a partnership, each enterprise has cooperative as well as competitive motives. The cooperative aspect arises from the fact that firms can collectively use their knowledge to produce something that is beneficial to them all (common benefits). The competitive aspect is a consequence of each firm's attempt to use the knowledge of its partners for private gains, the motive for setting up Strategic Networks (Hemphill and Vonortas, 2003, pp. 260–261). For a sustainable partnership, a combination of private and common benefits is needed, its ratio described by the relative scope (Khanna *et al.*, 1998, p. 195). When private benefits are the only motive of a company, racing behaviour will arise and the alliance will be cancelled after a while. Kale *et al.* (2000) demonstrate the same based on a contingency model for inter-organisational learning and opportunistic behaviour. Henceforth, the perception of agents in networks about the relative scope will drive their behaviour and ultimately the development of the network.

That imperative of perception (about the relative scope) is also found in the final concept of evolutionary biology: autopoiesis. This theory offered by Maturana and Varela (1980) explains some principles of evolutionary processes: the ability to self-create or self-renew through a closed system of relations. In this view, living systems engage in circular patterns of interaction whereby change in one element of the system is coupled with changes elsewhere, setting up continuous patterns of interaction that are always self-referential. A system enters only interactions that are specified by its organisation. Therefore, a system's interaction with its environment is ultimately a reflection and part of its own organisation. An autopoietic system interacts with its environment in a way that facilitates its own self-creation. The structural coupling governs by which interactions a component of living system is influenced. When interactions initiate changes in the structure and composition, the structure is called plastic. Through repeated interaction and initiations (Kay, 2001, p. 472), the selection of subsequent structures is affected by the environment. Hence, the selection is driven by the environment and by the plasticity of the structure by its own components and internal relationships (Danilovic and Winroth [2006] performed an exploratory case study into networks using these concepts). The environment does not determine the internal adaptations! Therefore, autopoietic systems are interactively open and structurally closed (van der Vaart, 2002, p. 11). Hence, the relative scope of agents in networks is determined by the internal structure and the interaction with the environment (the network is only part of the environment); according to Dougall (1999, p. 709) this applies also to the social domain.

The interaction with the environment, when we consider agents in a network, depends on cognition. One of the foremost reasons for the research into autopoiesis stems from the quest for the nature of perception and cognition (e.g. Hernes and Bakken [2003] on Luhmann's interpretation for social systems, which is not settled yet). Perception and cognition derive from the nerve system realised by the autopoiesis of the organism. To exist, interactions should be continuously repeated since structural coupling exists; in this sense cognition represents gathering knowledge about all effective interaction for sustainability (see Maula [2000, p. 161] for its application to the case of Arthur D. Little). Learning as a process of cognition originates in the properties of self-reference of the system (coined in the 1980s by de Geus [1999, p. 111], which inspired the movement of the Learning Organisation by Senge [1992]). When learning exceeds the level of direct interaction and moves towards orientation in the common domain of two autopoietic systems, communication is established; Moriarty and Miikkulainen (1998, p. 385) demonstrate that this benefits collaborative coevolution. When descriptions of communication and interaction lead an entity to become an observer of its own behaviour, self-conscience arises. The composition of a system related to an external point of reference defines the identity of an autopoietic system (van der Vaart, 2002, pp. 7, 24–25). The identity is strongly related to the composition of the entity, changes in the composition lead to a changed identity; through self-reference autopoietic systems seek to maintain their identity unless perturbations provoke adaptations. These notions led Mingers (1995) to connect the theory of autopoiesis to the systems hierarchy of Boulding (1956). But it also led to models for interaction between agents, where perception of the interaction drives the (mutual) behaviour of agents (e.g. Zeleny [2001] on the "Third Italy"); a comparative case study by Mota and de Castro (2004) confirms this position. Moreover, it tells that **information sharing** in industrial networks, driven by both perception and relative scope, drive the dynamic behaviour of individual agents, an area not yet researched explicitly; e.g. information sharing as a theme emerges since suppliers express themselves regularly about the lack of adequate information (both strategic and operational) from their customers in collaborations.

5.4 Outline for Interdisciplinary Research

The new paradigm for industrial networks requires the intense collaboration between the mentioned domains of natural science (in the previous section) and management science. This could form the base for research that will move away from existing approaches to collaboration; research should target the interdisciplinary development of a generic model of complexity as a basis for a problem-to-system match framework for collaborative systems in production industry. The core is the application of principles of complex systems theory (and related strands of research in evolutionary biology) from natural sciences to collaborative enterprise networks as socio-technical systems. These projects should attempt to understand these networks as complex systems that can only assure their viability through adaptation

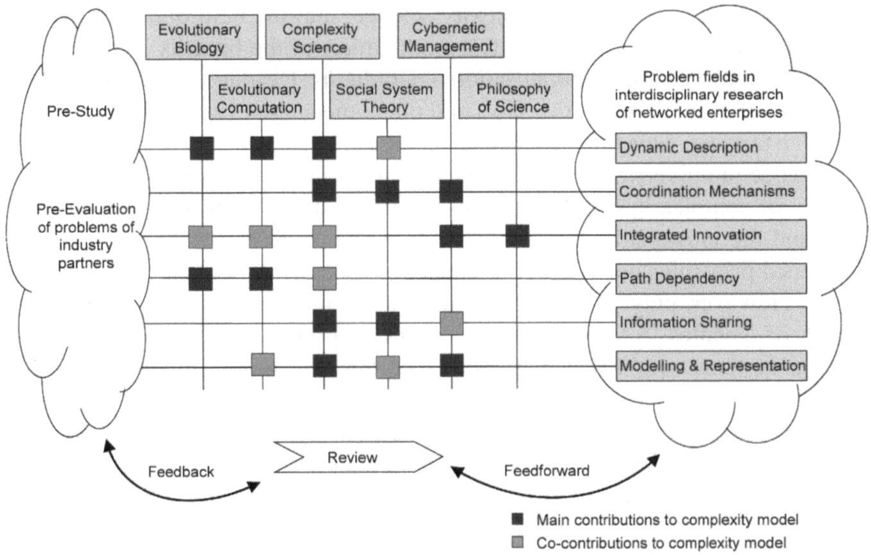

Figure 5.7. *Scientific pattern to resolve problem fields of enterprise networks via an interdisciplinary research approach (as proposed for a project in Dekkers et al. [2004]). The six themes (dynamic description, coordination mechanisms, integrated innovation, path dependency, information sharing, modelling & representation) stem from earlier research. The link has been made to the individual disciplines. Furthermore, the industry partners in this project contribute to linking the developments throughout the project to real case studies.*

in within these inter-organisational networks. Five themes have been identified based on the challenges for collaboration: dynamic description, coordination possibilities, integrative innovation, path dependency and information sharing (they have been highlighted in the text). Furthermore, modelling and representation constitutes a core of the approach that will be outlined below. The five themes serve as focal point to integrate the contributions of the individual disciplines in addition to the modelling and representation that arrives from deploying concepts of natural sciences to the challenges of industrial networks.

To address these themes, and to resolve industrial problems of networked organisations, a pre-evaluation of different scientific disciplines has been performed. Figure 5.7 gives an overview of the themes and the contribution of different scientific disciplines to the resolution of collaborative problems. Starting point are the identified problems of production industry. Mirrored to these, the development and adoption of scientific approaches is continuously reviewed to achieve major scientific progress in the six themes. The contribution of natural sciences will address the challenges of industrial networks and generate new theory. Contemporary literature in management science attributes the functioning and success and failure in collaboration to common pitfalls, such as culture, resistance to change, change management, working methods. By deploying theories from natural sciences to the research challenges of industrial networks, not only will new theory be generated, those causes for failures of collaborations might find new explanations.

The recent insights from complexity science, into the field of collaboration in production industry, will allow the full exploitation of the benefits it may yield in comparison with traditional organisational paradigms. Despite the great complexity and variety of systems, universal laws and phenomena are essential to their inquiry and understanding. Scientific endeavour is based, to a greater or lesser degree, on the existence of universality, which manifests itself in diverse ways. In this context, the study of complex systems as a new endeavour strives to increase the ability to understand the universality that arises when systems are highly complex. A study of universal principles does not replace a detailed description of particular complex systems. However, universal principles and tools guide and simplify inquiries into the study of specifics. A careful articulation of such principles enables us to approach particular systems with a systematic guidance that the studies of complex systems often lack.

For example, Agent-Based Modelling is a new and special branch of computer simulation that emerged as a methodology for studying complex systems (Buchanan, 2004). Agent-Based Models consist of agents that have states and behavioural rules, and an environment. In the environment, either spatial (e.g. a rectangular grid) or non-spatial (e.g. an abstract trading community), interactions among agents take place. The interactions can be direct, where the action immediately changes the state of the partner, or indirect, when the action changes the environment, which, in turn, causes the partner's state to change. Similarly, theoretical evolutionary biology has recently used game theories to explain and describe phenomena related to speciation. Especially, Adaptive Dynamics (Geritz *et al.*, 1997; Meszéna *et al.*, 2001) beholds new outlooks for describing stability in populations (Dekkers, 2005b), thereby relating the development of species to state spaces (a description of fitness landscapes [Caldart and Ricart, 2004]). Traditional social sciences, especially classical economics, have very strong assumptions about the rationality of agents. Most Agent-Based Modelling uses bounded rational agents that have only local, limited information, and limited ability and time to process that information, similar to real-life situation in industrial networks.

5.6 Concluding Remarks

Industrial networks give opportunities for individual companies to capture market opportunities more adequately. Yet, these networks require different approaches because of the characteristics of collaboration, decentralization of decision-making and inter-organisational integration. Especially, the move to more loosely connected agents in networks calls for new paradigms accounting for dynamics and complexity rather than adhering to the more static approaches for individual companies.

5.6.1 Implications for Research

With the field of complexity research still being a scattered patchwork of insights, a pragmatic and interdisciplinary approach holds the potential of yielding valuable insights into complexity modelling in today's networked (production) industry (e.g.

as advocated by Stacey [1993a]). It is the intention to stretch the notions beyond sometimes vague, rallying metaphors that yield at best new perspectives at known phenomena, as Wilson (1998, p. 88) captures the contribution of the science of complexity); the research explicitly aims at understanding patterns of collaboration in industrial networks. The most common approaches focus on the complexity of structures (mostly internal complexity) with a static character; this links to the most common system theories. In our view, the dynamic dimension of complexity, found in recent progress in natural sciences, will fit the characteristics of industrial networks.

Any undertaking to understand complexity in any imaginable system aims at making the world around us more accessible to our mental frameworks and thus predictable. With the need and ambition to systematically understand and influence artificial and socio-technical systems, any improvement in our notion of the world of complexity enables us to develop instruments and tools for wielding the potential gained through new insight. This means that with the extent of understanding what complexity is and how to appropriately describe it, the basis for new technological solutions is laid. The complexity models aim at making complexity management in socio-technical systems more accessible and more likely. The objectives of such endeavours is to provide a properly designed framework-of-thought for the (technological) implementation of complexity management infrastructures, which rely either on state of the art information technologies or on new insights about the architecture and characteristics of complex systems. Consequently, the objectives are to create a framework for complexity controlling systems for future network management tasks.

5.6.2 Industrial Implications

The implementation of this framework enables companies to react more flexibly when market opportunities arise, thus increasing their competitive position, and to manage the networks they participate in more adequately. The National Research Council (Bollinger, 1998, pp. 18, 131) stresses that manufacturing centres will operate in networks, and that intra-organisational and inter-organisational structures will be increasingly based on flexible, transient cooperation models. An approach based on the science of complexity and the science of evolutionary biology will contribute to this quest facing manufacturing companies, since it aims at understanding patterns of collaboration. The understanding of these patterns can be guided to methods and tools for industry and could be developed after this basic research.

The adaptation to changing environmental conditions and the drive for innovation and fast product development will benefit from the results of these proposed research projects. New paradigms for industrial networks will stretch beyond the traditional issues of trust, power, and supply chain management. Results should guide companies managing the dynamics of the networks, the higher degree of specialisation, the development and implementation of technologies, and the development of appropriate long or short-term relationships. This will reflect on both the optimization of the supply chains and the speed of innovation and product development.

Acknowledgements

This chapter has been based on two earlier papers (Dekkers *et al.*, 2004; 2005). The author would like to thank Günther Schuh, Alexander Sauer, Martin Schönung and Patrick Wegehaupt for their contributions to those papers that constitute the base of this chapter.

References

Ashby, W.R. (1956) An introduction to cybernetics, Wiley, New York.

Ashmos, D.P., Duchon, D. and McDaniel Jr., R.R. (2000) Organizational responses to complexity: the effect on organizational performance, Journal of Organizational Change Management, Vol. 13, No. 6, pp. 577–595.

Beaumont, N. and Khan, Z. (2005) A Taxonomy of Refereed Outsourcing Literature, 22/05, Monash University, Victoria, Australia.

Beer, S. (1959) Cybernetics and Management, Wiley, New York.

Beer, S. (1966) Decision and control; the meaning of operational research and management cybernetics, Wiley, London.

Beer, S. (1972) Brain of the Firm - the Managerial Cybernetics of Organization, Wiley, Chichester, UK.

Bierach, B. (2000) Kauf was Du kennst, Wirtschaftswoche, No. 6, pp. 144–152.

Bollinger, J.G. (1998) Visionary Manufacturing Challenges for 2020, National Academy Press, Washington.

Bonabeau, E. (1998) Social Insect Colonies as Complex Adaptive Systems, Ecosystems, Vol. 1, No. 5, pp. 437-443.

Bonabeau, E. and Meyer, C. (2001) "Swarm Intelligence: A Whole New Way to Think About Business", Harvard Business Review, Vol. 79, No. 5, pp. 107–114.

Boswijk, H.K. (1992) Complexiteit in evolutionair and organisatorisch perspectief, het zoeken naar balans tussen vermogens en uitdagingen, Erasmus Universiteits Drukkerij, Rotterdam.

Boulding, K.E. (1956) General Systems Theory, Management Science, Vol. 2, No. 3, pp. 197–208.

Bouvier-Patron, P. (2001) Analysis of Change Within and Between Firms embedded in their environment in the context of increasing interactions: understanding complexity of organizations, European Journal of Economic and Social Systems, Vol. 15, No. 1, pp. 1-38.

Buchanan, M. (2004) It's the Economy, stupid, New Scientist, Vol. 182, No. 2442, pp. 34–37.

Burnes, B. (2005) Complexity theories and organizational change, International Journal of Management Reviews, Vol. 7, No. 2, pp. 73–90.

Burns, T. and Stalker, G.M. (1961) The management of innovation, Tavistock Publications, London.

Caldart, A.A. and Ricart, J.E. (2004) Corporate Strategy Revisited: a View from Complexity Theory, European Management Review, Vol. 1, No. 1, pp. 96–104.

Camarinha-Matos, L.M. and Afsarmanesh, H. (2005) Collaborative Networks: a new scientific discipline, Journal of Intelligent Manufacturing, Vol. 16, No. 4–5, pp. 439–452.

Cánez, L.E., Platts, K.W. and Probert, D.R. (2000) Developing a framework for make-or-buy decisions, International Journal of Operations & Production Management, Vol. 20, No. 11, pp. 1313–1330.

Car, Z., Hatono, I. and Ueda, K. (2002) Layered Based Architecture for Adaptive Reconfiguration of Manufacturing System, in Proceedings of the 4th International Workshop on Emergent Synthesis, Kobe, 9–10 May.

Checkland, P. (1981) Systems Thinking, Systems Practice, Wiley, Chichester, UK.

Choi, T.Y., Dooley, K.J. and Rungtusanatham, M. (2001) Supply networks and complex adaptive systems: control versus emergence, Journal of Operations Management, Vol. 19, No. 3, pp. 351–366.

Colotla, I., Yongjiang, S. and Gregory, M.J. (2003) Operation and performance of international manufacturing networks, International Journal of Operations & Production Management, Vol. 23, No. 10, pp. 1184–1206.

Croom, S., Romano, P. and Giannakis, M. (2000) Supply chain management: an analytical framework for critical literature review, European Journal of Purchasing & Supply Management, Vol. 6, No. 1, pp. 67–83.

Danilovic, M. and Winroth, M. (2006) Corporate manufacturing network: from hierarchy to self-organising system, International Journal of Integrated Supply Management, Vol. 2, No. 1–2, pp. 106–131.

Dannenmaier, S., Lindebner, G., Saalfrank, S., Morschett, D., Neidhart, M. and Zentes, J. (2003) Die Zukunft der Kooperationen, Deutscher Fachverlag, Frankfurt.

Dekkers, R. (2000) Decision models for outsourcing and core competencies in manufacturing, International Journal of Production Research, Vol. 38, No. 7, pp. 4085–4096.

Dekkers, R. (2005a) Outsourcing Revisited, in Proceedings of the 18th International Conference on Production Research, Salerno, 31 July–4 August.

Dekkers, R. (2005b) (R)Evolution, Organizations and the Dynamics of the Environment, Springer, New York.

Dekkers, R. (2006) Engineering Management and the Order Entry Point, International Journal of Production Research, Vol. 44, No. 18–19, pp. 4011–4025.

Dekkers, R. (2007) Distributed Manufacturing as Co-Evolutionary System, in Proceedings of the 19th International Conference on Production Research, Valpairaso, 31 July–4 August.

Dekkers, R. and Luttervelt, C.A. van (2006) Industrial networks: capturing changeability?, International Journal of Networking and Virtual Organisations, Vol. 3, No. 1, pp. 1–24.

Dekkers, R., Sauer, A., Schönung, M. and Schuh, G. (2004) Collaborations as Complex Systems, in Proceedings of the Designing and Operating Global Manufacturing & Supply Networks, 9th Annual Cambridge International Manufacturing Symposium, IMNet/CIM, Cambridge, 25–26 September.

Dekkers, R., Sauer, A., Schönung, M. and Schuh, G. (2005) Collaborations as Complex Systems, in Proceedings of the International Conference on Changeable, Agile, Reconfigurable and Virtual Production, München, 22–23 September.

Dekkers, R. and Wood, K.L. (2007) Innovative Manufacturing Programme - Overall Mid-Term Review of the Innovative Manufacturing Research Centres, University of Paisley, Paisley.

Dougall, C. (2003) Autopoiesis and Aristotle: Rethinking organisation as form, Kybernetes, Vol. 28, No. 6/7, pp. 777–791.

Fischer, D. and Hafen, U. (1997) Immer wieder die gleichen Fehler, io Management Zeitschrift Industrielle Organisation, Vol. 66, No. 6, pp. 38–45.

Foerster, H.v. (1985) Einführung in den Konstruktivismus, R. Piper, München.

Fricker, A.R. (1996) Eine Methodik zur Modellierung, Analyse und Gestaltung komplexer Produktionsstrukturen, RWTH Aachen, Aachen.

Friedli, T. (2000) Die Architektur von Kooperationen, Universität St. Gallen, Difo: Bamberg.

Geritz, S.A.H., Metz, J.A.J., Kisdi, E. and Meszéna, G. (1997) Dynamics of Adaptation and Evolutionary Branching, Physical Review Letters, Vol. 78, No. 10, pp. 2024–2027.

Geus, A. de (1999) The Living Company, Nicolas Brealy, London.

Goldman, S.L. and Nagel, R.N. (1993) Management, technology and agility: the emergence of a new era in manufacturing, International Journal of Technology Management, Vol. 8, No. 1/2, pp. 18–38.

Guare, J. (1990) Six Degrees of Separation: A Play, Vintage Books, New York.

Gunasekaran, A., Macbeth, D.K. and Lamming, R. (2000) Modelling and analysis of supply chain management systems: an editorial overview, Journal of the Operational Research Society, Vol. 51, No. 10, pp. 1112–1115.

Hafeez, K., Zhang, Y. and Malak, N. (2002) Determining key capabilities of a firm using analytic hierarchy process, International Journal of Production Economics, Vol. 76, No. 1, pp. 39–91.

Hemphill, T.A. and Vonortas, N., S. (2003) "Strategic Research Partnerships: A Managerial Perspective", Technology Analysis & Strategic Management, Vol. 15, No. 2 pp. 255-271.

Hernes, T. and Bakken, T. (2003) Niklas Luhmann's Autopoiesis and Organization Theory, Organization Studies, Vol. 24, No. 9, pp. 1511–1535.

Houchin, K. and MacLean, D. (2005) Complexity Theory and Strategic Change: an Empirically Informed Critique, British Journal of Management, Vol. 16, No. 2, pp. 149–166.

Huggins, R. (2000) The success and failure of policy-implanted inter-firm network initiatives: motivations, processes and structure, Entrepreneurship & Regional Development, Vol. 12, No. 2, pp. 111–135.

Humphreys, P., McIvor, R. and Huang, G. (2002) An expert system for evaluating the make or buy decision, Computers & Industrial Engineering, Vol. 42, No. 2–4, pp. 567–585.

Jansen, S.A. and Petersen, J. (2000) Mythos "Merger-Misserfolg"? Prozedere, Probleme and Potentiale der Erfolgsmessung von Unternehmenszusammenschlüssen, M&A Review, No. 12, pp. 470–475.

Jost, J. (2004) External and internal complexity of complex adaptive systems, Theory in Biosciences, Vol. 123, No. 1, pp. 69–88.

Kale, P., Singh, H. and Perlmutter, H. (2000) Learning and Protection of Proprietary Assets in Strategic Alliances: Building Relational Capital, Strategic Management Journal, Vol. 21, No. 3, pp. 217–237.

Karlsson, C. (2003) The development of industrial networks: Challenges to operations management in an extraprise, International Journal of Operations & Production Management, Vol. 23, No. 1, pp. 44–61.

Kauffman, S. (1995) At home in the universe: the search for laws of self-organization and complexity, Oxford University Press, New York.

Kauffman, S.A. (1993) The Origins of Order: Self-Organization and Selection in Evolution, Oxford University Press, New York.

Kay, R. (2001) Are Organizations Autopoietic? A Call for New Debate, Systems Research and Behaviorial Science, Vol. 18, No. 6, pp. 461–477.

Kennedy, J.F., Eberhart, R.C. and Shi, Y. (2001) Swarm intelligence, Morgan Kaufmann, San Francisco.

Khanna, T., Gulati, R. and Nohria, N. (1998) The dynamics of learning alliances: competition, cooperation and relative scope, Strategic Management Journal, Vol. 19, No. 1, pp. 193–210.

Klemm, K., Eguíluz, V., Toral, R. and Miguel, M.S. (2003) Nonequilibrium transactions in complex networks: A model of social interaction, Physical Review E, Vol. 67, No. 2, pp. 026120-1/026120-6.

Krapivsky, P.L. and Redner, S. (2001) Organization of growing random networks, Physical Review E, Vol. 63, No. 6, pp. 066123/1-14.

Lacity, M.C. and Hirschheim, R.A. (1995) Beyond the information systems outsourcing bandwagon: the insourcing response, Wiley, Chichester, UK.

Lamming, R., Johnsen, T., Zheng, J. and Harland, C. (2000) An initial classification of supply networks, International Journal of Operations & Production Management, Vol. 20, No. 6, pp. 675–691.

Lewin, A.Y., Long, C.P. and Carroll, T.N. (1999) The Coevolution of New Organizational Forms, Organization Science, Vol. 10, No. 5, pp. 535–550.

Limerick, D., Crowther, F. and Cunnington, B. (2002) Managing the new organisation collaboration and sustainability in the postcorporate world, Allen & Unwin, Crows Nest, N.S.W.

Malik, F. (1992) Strategie des Managements komplexer Systeme, Haupt, Bern.

Maturana, F.P. and Norrie, D.H. (1996) Multi-agent Mediator architecture for distributed manufacturing", Journal of Intelligent Manufacturing, Vol. 7, No. 4, pp. 257–270.

Maturana, H.R. and Varela, F.J. (1980) Autopoiesis and Cognition, Reidl, London.

Maula, M. (2000) The senses and memory of a firm - implication of autopoiesis theory for knowledge management, Journal of Knowledge Management, Vol. 4, No. 2 pp. 157–161.

McKelvey, B. (1999) Complexity Theory in Organization Science: Seizing the Promise of Becoming the Fad? Emergence, Vol. 1, No. 1, pp. 5–32.

McKendrick, D.G. and Carroll, G.R. (2001) On the Genesis of Organizational Forms: Evidence from the Market for Disk Arrays, Organization Science, Vol. 12, No. 6, pp. 661–682.

Meszéna, G., Kisdi, É., Dieckmann, U., Geritz, S.A.H. and Metz, J.A.J. (2001) Evolutionary Optimisation Models and Matrix Games in the Unified Perspective of Adaptive Dynamics, Selection, Vol. 2, No. 1–2, pp. 193–210.

Micklethwait, J. and Wooldridge, A. (1996) The Witch Doctors - What the Management Gurus are Saying, Why it Matters and How to Make Sense of it, Heinemann, London.

Miles, R.E. and Snow, C.C. (1984) Fit, Failure and the Hall of Fame, California Management Review, Vol. 26, No. 3, pp. 11–28.

Milgram, S. (1967) The Small World Problem, Psychology Today, Vol. 2, pp. 60–67.

Mingers, J. (1989) An introduction to autopoiesis - Implications and applications, Systemic Practice and Action Research, Vol. 2, No. 2, pp. 159–180.

Mingers, J. (1995) Self-Producing Systems: Implications and Application of Autopoiesis, Plenum Press, New York.

Morgan, G. (1997) Images of organization, Sage Publications, Thousand Oaks, CA.

Moriarty, D.E. and Miikkulainen, R. (1998) Forming Neural Networks Through Efficient and Adaptive Coevolution, Evolutionary Computation, Vol. 5, No. 4, pp. 373–399.

Mota, J. and Castro, L.M. de (2004) A Capabilities Perspective on the Evolution of Firm Boundaries: A Comparatice Case Example from the Portugese Moulds Industry, Journal of Management Studies, Vol. 41, No. 2, pp. 295–316.

Nassimbeni, G. (1998) Network structures and co-ordination mechanisms: A taxanomy, International Journal of Operations & Production Management, Vol. 18, No. 6, pp. 538–554.

Nelson, R.R. and Winter, S.G. (1982) An Evolutionary Theory of Change, Belknap Press, Cambridge, MA.
Newman, M.E.J. (2003) The Structure and Function of Complex Networks, SIAM Review, Vol. 45, No. 2, pp. 167–256.
Nieuwstad, J.P.M. (1997) De organisatie als complex adaptief systeem, Section Industrial Organisation and Management/Delft University of Technology, Delft.
Noori, H. and Lee, W.B. (2002) Factory-on-demand and smart supply chains: the next challenge, International Journal of Manufacturing Technology and Management, Vol. 4, No. 5, pp. 372–383.
O'Neill, H. and Sackett, P. (1994) The Extended Manufacturing Enterprise Paradigm, Management Decision, Vol. 32, No. 8, pp. 42–49.
Perona, M. and Miragliotta, G. (2004) Complexity management and supply chain performance assessment. A field study and a conceptual framework, International Journal of Production Economics, Vol. 90, No. 1, pp. 103–115.
Probert, D. (1997) Developing a make or buy strategy for manufacturing business, Institution of Electrical Engineers, London.
Riedl, R. (2000) Strukturen der Komplexität: eine Morphologie des Erkennens und Erklärens, Springer, Berlin.
Romme, A.G.L. and Witteloostuijn, A.v. (1997) Autopoiesis, Chaos en Zelforganisatie in de Bedrijfskunde, Bedrijfskunde, Vol. 69, No. 2, pp. 63–1.
Ropohl, G. (1999) Philosophy of Socio-Technical Systems, Society for Philosophy and Technology, Vol. 4, No. 3, pp. 59–71.
Ryu, K. and Jung, M. (2003) Agent-based fractal architecture and modelling for developing distributed manufacturing systems, International Journal of Production Research, Vol. 41, No. 17, pp. 4233–4255.
Schuh, G., Millarg, K. and Göransson, A. (1998) Virtuelle Fabrik - Neue Marktchancen durch dynamische Netzwerke, Hanser Fachbuch, München.
Schuh, G. and Sauer, A. (2005a) Mit starken Partnern in die Zukunft. Entwicklungsnetzwerke sichern Wettbewerbsfähigkeit des Maschinenbaus, Zuliefermarkt, No. 9, pp. 3.
Schuh, G. and Sauer, A. (2005b) Chancen nutzen in volatilen Märkten - Die strategische Bedeutung von Entwicklungsnetzwerken für den mittelständischen Maschinenbau wächst, Scope, Vol. 9.
Schuh, G. and Wegehaupt, P. (2004) Die virtuelle Fabrik - Lessons learned zehn Jahre danach, in: Initiative für Beschäftigung OWL e.V./Universität Bielefeld/SURVEY GmbH/ Bertelsmann Stiftung (Hg.) (Ed.) Unternehmensnetzwerke. Fragen der Forschung - Fragen der Praxis, Kleine, Bielefeld, pp. 117–127.
Schuh, G. and Wegehaupt, P. (2004) Kooperation im Wandel: Collaborative Swarms als Antwort auf Diskontinuität, in: H. Luczak and V. Stich (Eds.) Betriebsorganisation im Unternehmen der Zukunft, Springer, Berlin, pp. 31–42.
Schwaninger, M. (2000) Das Modell Lebensfähiger Systeme: Ein Strukturmodell für organisationale Intelligenz, Lebensfähigkeit und Entwicklung, 35, Universität St. Gallen, St. Gallen.
Schwaninger, M. (2001) System theory and cybernetics, Kybernetes, Vol. 30, No. 9/10, pp. 1209–1222.
Senge, P.M. (1992) The Fifth Discipline, Century Business, Kent.
Sigmund, K. (1998) Complex Adaptive Systems and the Evolution of Reciprocation, Ecosystems, Vol. 1, No. 5, pp. 444–448.
Smith, A. (1776). Wealth of Nations, W. Strahan and T. Cadell, London.
Stacey, R. (1993) Strategy as order emerging from chaos, Long Range Planning, Vol. 26, No. 1, pp. 10–17.

Stacey, R.D. (1993) Strategic management and organisational dynamics, Pitman, London.
Stacey, R.D. (1996) Complexity and Creativity in Organizations, Berrett-Koehler, San Fransisco.
Stadtler, H. (2005) Supply chain management and advanced planning - basics, overview and challenges, European Journal of Operational Research, Vol. 163, No. 3, pp. 575–588.
Stock, G.N., Greis, N.P. and Kasarda, J.D. (1999) Logistics, strategy and structure: A conceptual framework, International Journal of Physical Distribution & Logistics, Vol. 29, No. 4, pp. 224–239.
Tucker, B., Furness, C., Olsen, J., McGuirl, J., Oztas, N. and Millhiser, W. (2003) Complex Social Systems: Rising Complexity in Business Environments, New England Complex Systems Institute, Cambridge, MA.
Ulrich, H. (2001) Systemorientiertes Management: das Werk von Hans Ulrich, Haupt, Bern.
Vaart, R. van der (2002) Autopoiesis! Zon of Onzin voor organisaties?, Delft University of Technology/Section Production Technology and Organisation, Delft.
Volberda, H.W. and Lewin, A.Y. (2003) Co-evolutionary Dynamics Within and Between Firms: From Evolution to Co-evolution, Journal of Management Studies, Vol. 40, No. 8, pp. 2111–2136.
Watts, D.J. and Strogatz, S.H. (1998) Collective dynamics of 'small-world' networks, Nature, Vol. 393, No. 6684, pp. 440–442.
Watzlawick, P. (1976) Wie wirklich ist die Wirklichkeit?: Wahn, Täuschung, Verstehen, R. Piper, München.
Wegehaupt, P. (2004) Führung von Produktionsnetzwerken, Dissertation Rheinisch-Westfälischen Technischen Hochschule Aachen, Aachen.
Wiendahl, H.-P. and Scholtissek, P. (1994) Management and Control of Complexity in Manufacturing, Annals of the CIRP, Vol. 43, No. 2, pp. 533–540.
Williamson, O.E. (1975) Markets and hierarchies, analysis and antitrust implications: a study in the economics of internal organization, Free Press, New York.
Wilson, E.O. (1998) Consilience: the unity of knowledge, Alfred A. Knopf, New York.
Wognum, P.M. and Faber, E.C.C. (2001) Infrastructures for collaboration in virtual organisations, International Journal of Networking and Virtual Organisations, Vol. 1, No. 1, pp. 32–54.
Wright, S. (1982) The shifting balance theory and macroevolution, Annual Review of Genetics, Vol. 16, pp. 1–19.
Zeleny, M. (2001) Autopoiesis (self-production) in SME networks, Human Systems Management, Vol. 20, No. 3, pp. 201–207.
Zhang, S. (2000). Electronic Business in China, in: Proceedings of the 10th International Forum on Technology Management, Danube, Vienna, 27 Nov.–1 Dec.

PART II:

Control and Coordination in Industrial Networks

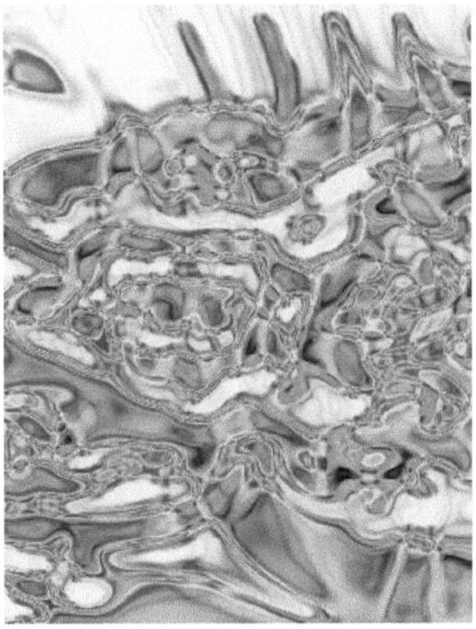

Complexity Perspective for Collaboration

The previous three chapters have explored the collaboration between loosely connected entities, all drawing on characteristics of complex systems to address issues of control and coordination. Those networks from loosely connected entities particularly suit collaboration between SMEs (Subsection 2.3.1 and Section 3.2), increasing their flexibility, improving their response time and yielding a wider variety of products. The applications of information and communication technologies deploy hubs for managing the business processes (Section 3.2, 3.7 and Section 4.4). Key to these hubs and managing the business processes is the capability of individual entities participating in these networks; called self-criticality by Kühnle in Section 4.2 and also appearing as dynamic internal adaptation in Section 3.5, that capability facilitates learning of the network and adaptation to changing circumstances. This capability strongly resembles the concept of process capability in the steady-state model that is mentioned by Dekkers (2005, p. 431). That then calls for reconfiguration, either by self-similarity based on fractals (Section 4.4) or by optimisation of the organelle structure (Subsection 2.3.1); note that the basis for these reconfiguration approaches – the integration of business processes: material flow and information flows – is the same. But above all, the collaboration calls for underpinning theory for co-evolution. That extends to the understanding of the role of private and common benefits (Chapter 5 provides an expansion of the thinking in the Appendix to Chapter 2), the dynamics behaviour resulting from interlinks between entities in the networks and possibily described by game-theories, Agent-Based Modelling, the *NK[C]* model, etc. (see Chapter 5). However, all of these notions provide a keystone to understanding collaboration in industrial networks that should constitute the basis for control and coordination in industrial networks.

Introducing Part II: Control and Coordination in Industrial Networks

Although Chapters 3 and 4 have already addressed the decentralised organisational decision-making and inter-organisational integration for coordination, the next two contributions in Part II elaborate in more detail on possible solutions for the coordination and control processes in industrial networks. In the context of Dispersed Manufacturing Networks, very few works or maybe none at all have been devoted to control and coordination in loosely connected networks. Freitag and Winkler (2000) hint at control mechanisms for networks and their implications but do not provide solutions. And Honavar and Uhr (1990) elaborate on coordination and control structures for networks but do not extend that to the domain of social networks, e.g. industrial networks. Choi *et al.* (2001) elaborate on chaos and control for coordination in supply chains; these chains can hardly be considered as networks of loosely connected entities. For supply chains, Li *et al.* (2007) demonstrate the instability of these networks with their simulation study. The next two chapters expand on agility as prerequisite for the responsiveness of Dispersed Manufacturing Networks. Petri *et al.* describe in Chapter 6 an open source solution for managing the logistics between firms in an industrial network with loosely connected entities.

Their approach supports the sustainability of Dispersed Manufacturing Networks, through the possibility to connect firms without (severe) interventions in internal systems and processes. In Chapter 7, Hossein Sharifi *et al.* present an integrated approach to facilitate dynamic and simultaneous design and development of products and supply chain contributing to the agility of the supply chain. This should enhance existing practices and approaches to product development process as well as supply chain development and management. Putting it all together, the two contributions to this book offer solutions to improve and to facilitate improved coordination and control in industrial networks with interdependent agents.

References

Choi, T.Y., Dooley, K.J. and Rungtusanatham, M. (2001) Supply networks and complex adaptive systems: control versus emergence, Journal of Operations Management, Vol. 19, No. 3, pp. 351-366.

Dekkers, R. (2005) (R)Evolution, Organizations and the Dynamics of the Environment, Springer, New York.

Freitag, M. and Winkler, I. (2000) Mechanisms of Coordination in Regional Networks, in Proceedings of the 7th International Conference on Multi-Organizational Partnerships and Cooperative Strategy, Leuven, 6–8 July.

Honavar, V. and Uhr, L. (1990) Coordination and control structures and processes: possibilities for connectionist networks (CN), Journal of Experimental & Theoretical Artificial Intelligence, Vol. 2, No. 2, pp. 277–302.

Li, G., Sun, L., Gu, Y. and Dong, Y. (2006) Self-organisation evolution of supply networks, International Journal of Manufacturing Technology and Management, Vol. 10, No. 2–3, pp. 142–160.

6 Designing and Modeling Agile Supply-Demand Networks

Petri Helo
Natalia Kitaygorodskaya
Sari Salminen
University of Vaasa

Roger J. Jiao
Georgia Institute of Technology

Abstract Online, on-demand and real-time availability of information to all members of a manufacturing system enables them to be agile and in the best position to react quickly, efficiently, synchronously, and collectively to the changing market. This chapter proposes a design and modeling tool for Agile Supply-Demand Network (ASDN) design. The software is open source and freely available for research and commercial uses. It supports modeling, analyzing and limited optimizing of supply-demand networks. Also the network level logistics analysis that is behind the modeling tool is discussed.

Keywords Agile manufacturing, Global manufacturing, Logistics information systems, Supply-demand networks

6.1 Introduction

"Supply-demand network" is a relatively new term in the field of supply chain management. It stresses the fact that relationships between different supply, manufacturing and distribution units are more complex than in a supply chain. The supply-demand network perspective has evolved out of two distinct streams: descriptive research on industrial networks, which has provided a general understanding of complex buyer-supplier relationships, and prescriptive research on supply chain management. The latter has been evolving gradually since the 1960s by increasing the scope of academic and managerial attention from a company's internal chain, through dyadic and second-tier relationships, to supply chains in the late 1980s and recently to supply-demand networks. The supply-demand network perspective reflects a general shift in supply chain management from operational to strategic decision-making (Harland *et al.*, 2001). The strategic level of supply chain management includes decisions on the number and location of companies in each tier, choice of transportation channels, and management of information along the supply-demand network. It has been recognized by practitioners that a great share of supply chain costs are locked in supply-demand network design. Therefore, it is necessary to pay more attention to strategic decision-making regarding the supply-demand network structure.

According to the design school of strategic management, the organizational structure should follow the strategy. It seems it will not be a mistake to apply the same logic to supply chain management and say that the supply-demand network structure should follow the company's supply chain strategy. There are many papers that focus on either supply chain strategies or supply-demand network structures. The former usually focus on a specific type of strategy. The latter present models developed mostly in operational research under the rubric of supply-demand network design. Those models seek to find the optimal (or nearly optimal) solution for an objective function and rather facilitate supply-demand network strategic decision-making than discuss supply-demand network structure per se. A coherent discussion of supply-demand network structures is missing in the literature.

The scope of this chapter is to clarify the current situation of the supply-demand network by presenting a modeling tool that helps determining customer demand, inventory levels, lead-times, and routes. A typical industry problem could be stated as: *How to develop lead-time and reduce inventory levels in complex global supply-demand networks?* This problem may be broken down into the following sub-problems:

- Which node has the biggest impact on the supply-demand network in terms of improvement potential?
- How can the performance of the chain be optimized based on customer demand and service level?
- How can the improvement be measured and verified in financial terms?

In order to answer these questions, an open source software tool is presented in this chapter. The software helps to analyze the structure and performance of dispersed networks and estimate the value of improvement potential. This chapter presents an overview and puts together several supply chain strategies and their decision determinants. Then a discussion of different structures that facilitate implementation of the strategies follows.

6.2 Supply Chain Strategies and Their Decision Determinants

According to Hicks (1999, p. 26) the aim of strategic planning in supply chain management is "to arrive at the most efficient, highly *profitable* supply chain system that *serves customers in a market*". Some markets exhibit similar properties. For example, on some markets the demand is stable, whereas on others it is volatile. On some markets customers value low costs, but on others they prefer a high service level. This grouping suggests that markets with similar properties may require similar strategies. These generic strategies are discussed below.

Supply chain management is the management of five processes that link companies in supply-demand networks: plan, source, make, deliver, and return (Supply Chain Council, 2008). Two groups of supply chain management strategies are recognized: postponement/speculation and leanness/agility strategies (Helo, 2006).

The concept of postponement has been discussed by many authors (Naylor *et al.*, 1999). Manufacturing postponement is the postponement of product differentiation to the latest possible point (Pagh and Cooper, 1998). This means that a final product assembly is moved downstream a value-added network, closer to the point of purchase. The concept of postponement is closely intertwined with that of modularization (Ernst and Kamrad, 2000). Modularization implies that products in a certain product family are designed in such a way that all of them consist of different standardized units. Combinations of these units produce different end products. Modularization facilitates postponement because it allows for production of standardized units in large volumes and their transportation downstream (e.g. to a distribution centre). Final assembly of the end product (or packaging and/or labelling) is performed only when the exact customer order is known. Different forms of manufacturing postponement, depending on at which point of the value-added chain final manufacturing is performed, are called make-to-order (closer to a point-of-sale), assemble-to-order strategies (closer to a manufacturing company), and engineering-to-order (on site of a manufacturing company).

Apart from manufacturing postponement, logistics postponement is also recognized (Pagh and Cooper, 1998). This is a postponement of downstream movements of products. To achieve this, an anticipatory inventory is maintained at one or few strategic locations. Logistics postponement is characterized by centralized inventories and direct distribution. A concept opposite to postponement is speculation (Pagh and Cooper, 1998). Speculation can also be manufacturing and/or logistics. It assumes that product differentiation and/or downstream movements of goods are performed at the earliest possible point. Manufacturing and logistics speculation and postponement can be used in all possible combinations and that results in four different strategies (Pagh and Cooper, 1998): manufacturing postponement strategy (manufacturing postponement combined with logistics speculation), logistics postponement strategy (manufacturing speculation combined with logistics postponement), full postponement strategy (manufacturing postponement combined with logistics postponement), and full speculation strategy (manufacturing speculation combined with logistics speculation).

Another group of supply chain strategies consists of leanness and agile strategies. Lean thinking has gained significant attention among organizations as a paradigm that serves to optimize performance and to improve competitive positions. Leanness aims at minimization of waste (*muda*) throughout the whole chain (Womack and Jones, 1996). During the 1990s and in the beginning of the 21st century, the agile manufacturing paradigm evolved as an alternative to lean thinking (Richards, 1996). It seeks to provide quick responses to changing market needs. While the main focus in leanness strategy is to minimize total costs, the agility strategy aims at providing high service level (Helo, 2004). Since the high service level cannot always be achieved when costs are minimized, these two strategies are often seen as alternative. However, increasingly, customers want not only a high service level but low costs as well. To achieve this, leanness and agility strategies can be combined into a hybrid strategy (Christopher and Towill, 2001).

One of the approaches to a hybrid strategy (Christopher and Towill, 2001) is closely connected to the subsequent decisions on the supply-demand network structure. That is to use lean methods until the order decoupling point and agile methods after it. The order decoupling point separates the part of supply chain based on planning and forecasted demand from the part based on customer orders and actual demand. This idea is similar to that of manufacturing postponement: to hold inventory in a modularized form at some point of a value-added network and complete the final product assembly only when a precise customer order is known.

A summary of supply chain strategies is presented in Table 6.1. As the table illustrates, the desired outcomes of different strategies are quite similar, whereas the methods used to achieve the goals vary.

6.2.1 Decision Determinants

The decision which strategy to choose should be based on how well a certain strategy satisfies customer needs. Market requirements that have to be met as well as other important characteristics that are necessary to take into account are called decision determinants.

According to Christopher and Towill (2001), two important market determinants are the nature of market demand (stable/volatile) and the nature of customer requirements (highly/low customized). Markets with a stable demand and a low level of customization can be successfully satisfied with a lean type of supply chain, and highly customized markets with an unstable demand are better served when a supply chain is agile. If a final product should not only be available (high service level) but also affordable (low cost), a hybrid strategy is to be used.

Vonderembse *et al.* (2006) recognize other market determinants. These are the product characteristics and the stage of the product life cycle. Products can be standard (design changes incrementally; demand can be forecasted accurately; long life cycles), innovative (new products that satisfy emerging customer needs; demand cannot be forecasted; shorter life cycles) and hybrid (improvements are introduced periodically; long life cycles). The stages of the product life cycle are introduction, growth, maturity, and decline. Vonderembse *et al.* (2006) say that for standard products the lean supply chain strategy is the most suited. For innovative products during introduction and growth stages it is better to use the agility strategy, whereas during maturity and decline stages a lean or a hybrid strategy is more appropriate.

Table 6.1. *Desired outcomes and means of leanness/agility and postponement/speculation strategies*

Strategy	Desired outcome	Means
Speculation	High service level combined with reduced costs	Economies of scale, minimized number of stock outs
Postponement	High service level combined with low costs	Manufacturing and logistics postponement, mass customization
Leanness	Low costs	Waste elimination
Agility	High service level	Quick response to market needs

For hybrid products, the hybrid supply chain strategy is good throughout all product life cycle stages.

Among the decision determinants that help to choose between postponement and speculation strategies, the folllowing are similar to the leanness/agility decision determinants: market characteristics (demand, relative delivery time) and product characteristics (life cycle, type, range). Additionally, manufacturing and logistics characteristics should also be taken into account. For example, special manufacturing requirements may not allow postponement (Pagh and Cooper, 1998). Supply chain strategies and their decision determinants are presented in Table 6.2.

6.2.2 Inventory Location Within the Network

Inventories consist of raw materials, work in progress and finished goods, and they are used to help production or to satisfy customer needs. Stocks are also used as buffers between different demand and supply rates. The key reasons to carry inventories at different stages include:

1. *Protection against uncertainties and improved customer service.* Inventory systems face uncertainties in supply, future demand, and lead-times. Against these uncertainties companies carry safety stock, which might contain raw material, work in progress or finished goods. Raw materials are held in inventory to avoid stock-outs in the situations where suppliers have problems with delivery or quality of materials or components. Work in progress inventories are maintained because of poor maintenance, unreliable workers or fast schedule changes. Usually, better coordination of suppliers and customers in the supply-demand network helps to reduce safety stocks. Buffer stocks operate like safety stocks and they are often required between stages in the supply-demand network. Transit times influence buffer stocks; the longer the transit time, the more buffer stocks must be held as protection.
2. *Cost-effective production and purchase.* It is often economical to produce or purchase materials in lots. Producing in lots enables to spread the setup costs of the production machines over a large number of items. It also makes possible to use the same production equipment for different products. Sometimes it is

Table 6.2. *Decision determinants of supply chain strategies*

Decision determinants	Supply chain strategy
Stable demand, low customization, standard product or innovative product during maturity and decline stages of its lifecycle; special manufacturing capabilities; global company moves to local market	Lean; full speculation strategy
Volatile demand, high level of customization, innovative product during introduction and growth stages; manufacturing and logistics characteristics; current state of a company's position on a market	Agile; manufacturing postponement; logistics postponement; full postponement strategy
Hybrid product, innovative product during maturity and decline stages of its lifecycle; manufacturing and logistics characteristics; current state of a company's position on a market	Hybrid; manufacturing postponement; logistics postponement; full postponement strategy

also economical to purchase in large lots because of ordering costs, quantity discounts and transportation costs even though part of the items have to be held in inventory for later use. These kind of inventories are called cycle stocks.

3. *To provide for anticipated changes in demand or supply*. In the case of demand change, companies need to make decisions related to sourcing and capacity management. An example of such a situation may be a drastic change in the price or availability of raw materials or planned market promotion or if the business is seasonal.
4. *To cover transit times*. Transit inventories are goods that are on their way from one point to another or are waiting for transportation. The location of plants or warehouses and the mode of transportation affect inventory levels. The form of transport, transport routing and scheduling can control these inventories and affect transit times (Gattorna and Walters, 1996; Schary and Skjøtt-Larsen, 1995).

Inventories and inventory management have a huge impact on capital requirements, costs, and customer service. Inventory has a lower value at earlier stages of production and a higher value as products move closer to the customers. This also concerns the variety of inventories. If the amount of stock is based only on the inventory holding cost, it is preferable to hold it at earlier stages. Regardless the inventory form, companies want to release the capital tied up in stocks and to reduce, at the same time, the holding cost of these inventories. This can also improve the flexibility and responsiveness. Usually, material and component inventories are flexible because they are not yet tied up with finished products. However, if demand changes, finished goods require reworking, scrapping, and transportation from one location to another. Complete flexibility would eliminate inventories throughout the whole supply-demand network but this is only possible in situations where the stages from raw material production to customer delivery can be completely synchronized. In those supply-demand networks where products can be made directly to customer order, only the transit inventory between factory and customer is needed. Other inventories can be eliminated.

The inability to make accurate customer demand forecasts increases the variability and uncertainty in supply-demand networks. Almost all operational units suffer from these problems. Increased inventories obscure the transparency of final demand and cause a financial burden. Forecasting component demand is much easier than forecasting the demand of finished goods, because component demand is more stable especially when components can be used in various product families (Hadley, 2004; Schary and Skjøtt-Larsen, 1995).

If the inventory levels throughout the supply-demand network are reduced, operations will become more effective (Gattorna and Walters, 1996). In the past, inventories were located close to markets because customers preferred products with a short order lead-time. Nowadays, it is perceived that these inventories are unnecessary if the demand can be met through better coordination. If products can be made based on actual demand, they can flow without interruption directly to the end customer and there is no need for intermediate inventories. In this process the information system should be designed to reduce information delays and uncertainties.

When information and production systems are sufficiently responsive only one stocking point is needed as a buffer within a distribution system. If companies in a supply-demand network can develop relationships, advance demand visibility, centralize control, improve responsiveness of manufacturing and continuity of supply, and use flexible transportation systems to reduce delays in the flow of products, they have a huge opportunity to eliminate unnecessary stocking points (Hines *et al.*, 2000).

Currently, many companies are closing national warehouses and establishing distribution centres. This decision is based on the risk pooling effect, which means that consolidating inventories into fewer locations can substantially reduce total inventory requirement. A regional distribution centre serves a much wider geographical area than a national warehouse. It has been calculated that the reduction in the total inventory system that can be expected is proportional to the square root of the number of stock locations before and after rationalization. But it should be noted that centralized warehouses cause higher transportation costs because greater distances and airfreight might be needed to ensure short lead-times (Christopher, 1998).

6.2.3 Lead-Times

Managing lead-time in supply-demand networks is extremely important because it determines both cost and the ability to respond rapidly to the changing market needs (Schary and Skjøtt-Larsen, 1995). Traditionally, lead-time is defined as the time passed from receiving a customer order to the delivery of finished goods. Nowadays, a wider perspective is recommended. Lead-time could be defined as the elapsed period from product development, through purchasing, manufacturing, and assembly, to the delivery of the finished product. The management of this time span is particularly important to the success of the whole supply-demand network.

Lead-time has several constituents. As Karmakar (1993) notes: “The components of lead-times in manufacturing facility are queue time, processing time, and move or transportation time. The queues include waiting for move operations as well as queues for non-production tasks such as order processing, quality control and document generation. It is especially important to include the time from the instant the order is triggered, to the point that it is actually started into production. This time may involve no physical material in queue, but does have an impact on the lead-time experienced at the finished goods or customer stage. Hence, while it may not affect WIP, it can affect finished goods safety stock, and may also affect raw material inventory, or vendor inventory requirements”.

Usually supply-demand networks have some material or component supplies with long lead-times. Even though they react immediately to customer needs, they have to carry out requirements over long lead-time periods in order to have material at hand when orders arrive. The target for every organization should be to reduce lead-times at every stage in the supply-demand network as close as possible to zero, although zero lead-times hardly exist because every product has to be manufactured.

In many cases it is possible to find opportunities to reduce total lead-time. Often these opportunities can be implemented by quite simple changes in the systems. To find these opportunities, different stages in the supply-demand network have to be checked to see how the whole lead-time can be reduced through redesigning the way the chain is organized.

Transparency should increase when the speed of movement through the pipeline increases. In other words, every member of the supply-demand network should know where the product is at any moment, regardless the product ownership. This makes time an important resource, and the concept of strategic lead-time management is based on this assumption. Strategic lead-time management indicates that it is more effective to measure the total cumulative impact of the time used in the whole chain rather than time used by individual member of the chain. Customer operations can also be included in the management of time.

Implementation of time management should eliminate time wasting delays, decrease accumulation of quantities for batch processing, and provide immediate and accurate information for precise coordination. Changing the product's design or moving it closer to the customer adds product value. In the supply-demand network every unnecessary delay or unnecessary stage in production, like duplicated quality controls, are waste and they increase production costs.

6.2.4 Service Level or On-Time Delivery

The service level or on-time delivery is a percentage of orders that can be completely filled without any special processes. Special processes, like airfreight, can improve service level but 100% is extremely difficult to achieve. Christopher (1998) has an example how product availability effects service level: "If there are ten items on a particular order and each item is carried in stock at the 95% level of availability then the probability that the complete order can be filled is (0.95)10 which is 0.599. In other words, just over a 50/50 chance that we can satisfy the complete order". Hines *et al.* (2000) have introduced another way to calculate customer service in the single unit. The calculation is based on multiplication of the quality and delivery performance of each department. "If the quality performance of the purchasing department is 90 per cent and delivery performance of the suppliers is 50 per cent then the overall effectiveness of the inbound materials process is only 45 per cent. Faced with this situation, the purchasing department would concentrate its "de-bottlenecking" effort on the improvement of supplier delivery performance. Only when the combined quality and delivery performance of the supply base is near 100 per cent can costs be reduced by eliminating levels of inventory and such like".

6.3 ASDN Software Tool

Agile Supply-Demand Networks (ASDN) is an open source software package for designing industrial logistics networks in a user-friendly environment. The ASDN is implemented in Java programming language and the network files are stored as XML-files. The system can be integrated with other logistics systems by using this interface (Helo *et al.*, 2006). By changing the supply-demand network structure in the simulation model, the user can improve lead-time, agility and optimize inventory levels.

6.3.1 Notation of Models

Figure 6.1 shows the user interface of the ASDN software. There are two classes of network elements: nodes and arrows. Nodes are used to generate, modify, combine, and display participants in the network. Arrows represent supplier-customer relationships between the nodes. Below the attributes of nodes and arrows are given in more detail.

Node Input Attributes

1. General attributes:
   - ID: ordinal number of the node.
   - Label: name of the node.
   - Type: type of the node. The following types are possible: manufacturing, engineering, supplier, warehouse, sales company, end customer, distributor, wholesale, retailer.
   - Order Decoupling Point: Make-to-Stock; Assembly-to-Order; Make-to-Order; Engineer-to-Order; Capacity-Selling.
2. Finance:
   - Price: price of product per unit.
   - Cost per unit.

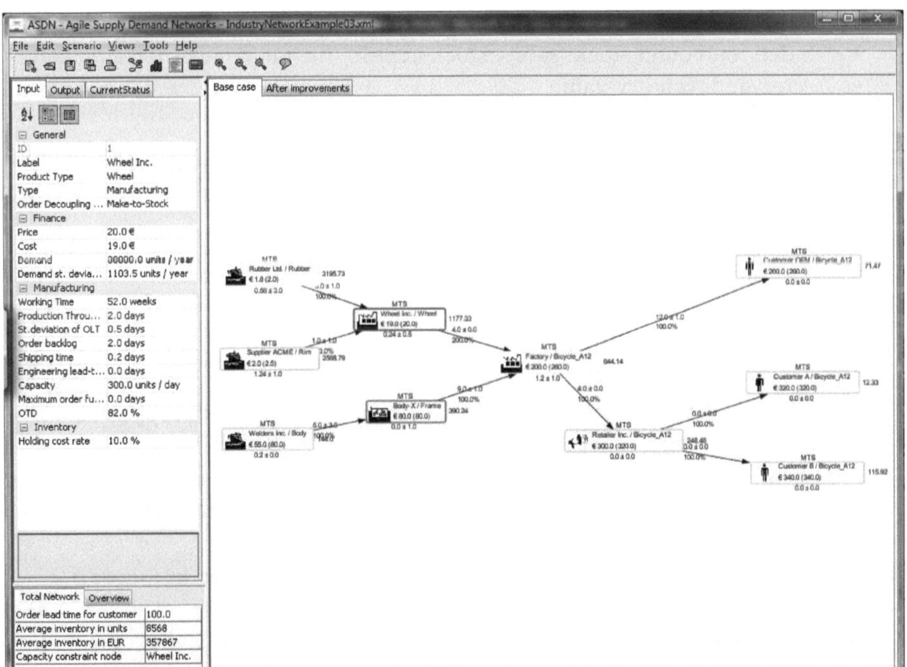

Figure 6.1. *User interface of ASDN software*

- Demand: demand per time period (aggregated from customer nodes).
- Demand standard deviation.

3. Manufacturing:
   - Working time in weeks.
   - Manufacturing stage time: time from beginning of production until the shipping of the product.
   - Standard deviation of manufacturing stage time.
   - Capacity per day.
   - Maximum order fulfillment time: maximum order fulfillment time asked by a customer (used as constraint in optimization).
   - Service level (OTD): percentage of how many orders have to be delivered on time.

For each node, a set of output variables is calculated based on input data. The output nodes calculations are also parameters, which can be edited by modifying the Java code in the user interface. Node output attributes are the following:

1. Manufacturing:
   - Order lead-time for manufacturing unit: time from the moment an order comes to the manufacturing unit until its fulfillment and product shipment to the next production unit or to the customer.
2. Inventory
   - Inventory. Average inventory in units. Cycle stock + Safety outbound stock + Safety inbound stock.
   - Cycle stock: amount of units required for the reorder time.
   - Safety inbound stock: safety stock kept for the uncertainty from supply.
   - Safety outbound stock: safety stock kept for the uncertainty from demand.
   - Average inventory value.
   - Inbound inventory value.
   - Inventory turn rate: demand per year divided by average inventory in units = Demand * 365 / Average inventory in units.
   - Inbound inventory value.
   - Outbound inventory value.

Arrow Attributes:

- ID: ordinal number of the arrow.
- Label: name of the arrow.
- Type: type of the transportation. The following types are possible: by ground, by air, by train, by ship, other.
- Freight attributes: cost, lead-time, lead-time C.V, service level (OTD), Current inventory level, minimum inventory level, maximum inventory level, lot size, weight.

Time definitions used in ASDN software are graphically presented in Figure 6.2.

A network is a combination of nodes and arrows that represents a layout of supplier-customer relationships between companies. The structure and performance are connected in network level analysis. For each network, the ASDN user can create

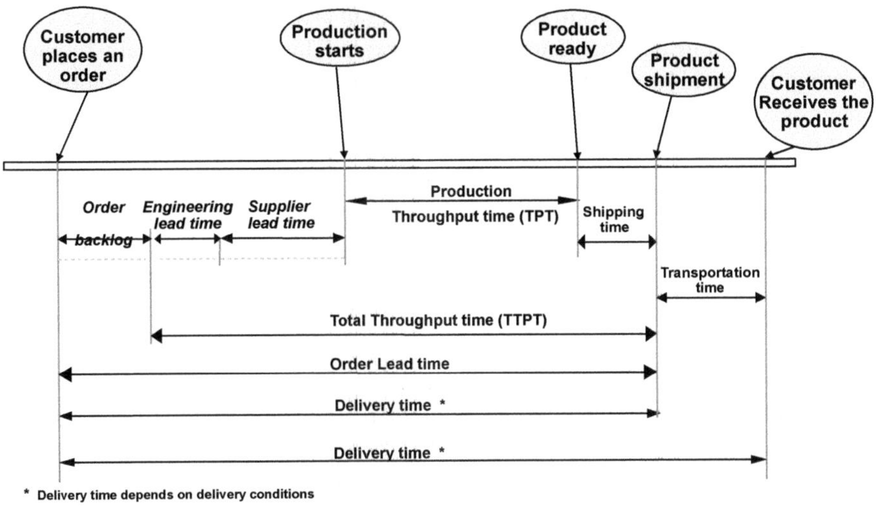

Figure 6.2. *Lead-time parameters*

several scenarios, which correspond to a certain layout of supplier-customer relationships between companies and/or values of companies' attributes. The comparison analyses typically include a base case scenario, which is assessed against potential improved scenarios. The ASDN software supports rapid modeling of large-scale logistics networks, where product information is given at the aggregated product family level.

6.3.2 Connection Between Lead-Times, Inventories and Service Level

A typical question in logistics analysis is to set objectives related to entire supply-demand network performance. The constraints include lead-time, inventory, and service levels – the parameters that have interdependencies. An example of questions behind development could be: How fast and effectively can customers' orders be fulfilled without affecting company profitability? Time lags in order handling increase overall distribution costs because larger inventories and faster transportation are needed to maintain customer service levels and cover time lags. Larger inventories and faster transportation will also increase working capital requirements and operating costs, but effective communications and information systems can eliminate some of these costs (Gattorna and Walters, 1996).

The centralization of global production in a few manufacturing sites may contradict the demands of the various markets and might also require local product variation. Because of the misguided assumption, manufacturing management might try to impose long lead-times to create buffer against the demands of different customers. As it was mentioned earlier, long lead-times do not protect against the uncertainty and fluctuation of demand, rather they make the situation worse. The determination of long manufacturing lead-times is usually an artificial constraint and in many cases it should be feasible to make order on a shorter span compared with the supply from stock, especially when customers are very special (Christopher, 1998).

An ordinary company has to carry inventories in order to fill in the gap between the logistics lead-time and the customer's order cycle. This situation usually requires forecasting. It should be remembered that no matter how sophisticated the forecasts are, their accuracy cannot be perfect and all mistakes in forecasting will end up as the problems of inventories (Christopher, 1998).

In a global chain it is usually essential to hold an intermediate inventory between manufacturing and customers. The purpose of these inventories is to buffer against extended transit times. But if the buffer's size is a sign of inflexibility in manufacturing or unsuccessful material management, then the need for equipment and stock holding in specific markets might well be questionable (Christopher, 1998).

Buffer inventories are required in situations where demands are unstable, lead-times are long, capacity constraints exist or there are some material, component or other constraints. It should be noted that the expected lead-time and lead-time variation determine safety stock levels for finished goods in pull systems. Because of this the time compression, management is very important, which takes on different meaning in specific situations (Karmakar, 1993; Schary and Skjøtt-Larsen, 1995).

Customer service can be improved by shorter order cycles and the costs of processing and inventories can be reduced by shorter throughput time. That also brings the whole chain closer to the end users. If the processing time can be decreased it reduces the need for forecasted orders and this helps to respond to orders already in the systems. Shorter processing time improves reaction times to the demand changes. It also improves flexibility in the changeovers of product and volume.

Less inventory results in less time and improved flexibility in manufacturing. The Just-In-Time (JIT) philosophy is an example of efficient production and it enables a flexible supply-demand network. The benefits of JIT are inventory reduction itself, a improved production process and better management in general.

Improved placement and deployment more often leads to situations where the right product is in the right place at the right time. This has both direct and indirect benefits. The direct benefit is an improved service level. The service level will improve because there are fewer stock-outs. If stock-outs appear, one of two things occurs. With luck, the customer will put off the purchase until the product is in stock. If not, the customer will find another source or give up buying the product. Both situations result in lost sales. The stock-out rate that leads to lost sales depends on the products and industries. Fewer stock-outs also mean lower transportation costs. Companies often transfer their stocks from one place to another or they serve their customers from different places to maintain their customer service levels in stock-out situations, also shipments have to accelerate to meet the customer demand. But these options increase overall supply-demand network costs (Hadley, 2004).

If a perfect match between the logistics lead-time and the customer's requirements is achieved there is no need for forecasts and inventories (Christopher, 1998). How well time is managed determines location and level of inventory and also determines the linking structure of the supply-demand network. Usually the measuring of time starts from the moment when the customer has a need for a specific product unit.

Preventing delays help to reduce inventories and especially to reduce throughput time (Schary and Skjøtt-Larsen, 1995).

Time is one of key variables in the supply-demand network. Therefore, time and the supply-demand network structure should be considered from two points of view. First, it should be thought of how transit time effects location of manufacturing stages. If transit times are long inventories should be bigger and this will increase inventory costs. So transit time can easily be converted into inventory cost. Long distances will require either long transit time by sea or expensive but fast air freight. When transit time is long, inventories will easily increase and become inflexible. These problems lead to another point that changing demand will be difficult to meet and ability of supply-demand network to respond to changes will be very slow.

There are different functional activities that the global dimension emphasizes. The emphasis varies between activities with specific circumstances. Preconditions for reduced inventories are decreasing (decreased) lead-times, minimizing (minimized) shipping quantities, and controlling (very careful control over) product movements. These tasks are more critical when transit times become longer. Inside Europe, transit times are shorter and because of that the emphasis turns to production and especially to minimization of production lead-times and increase in flexibility of production changes (Schary and Skjøtt-Larsen, 1995).

In Europe, the lengthy shipment times from/to the Far East, Australia, North, and South America are normal. The transit time from Japan to Rotterdam by sea is about five weeks. In comparison, the total length of time from consignment to arrival of air freight takes about five days which is approximately seven times shorter. The use of sea freight leads to (results in) remarkable investment in inventories on the high seas. It will also cause a serious problem with the use of the basic logistics principle of postponement, for example making the decisions of shipment at the last possible moment. The use of air freight grows when companies realize the composition of true supply-demand network costs. Penalties of high inventories and inflexible response to demands of markets should be taken into consideration. They can be avoided by shorter transit time and faster transit modes, but these alternatives will increase the supply-demand network costs. Delays and variability of lead-times in global supply-demand networks are diminished by shipping, consolidation, and customer clearance. These are major issues for most companies operating globally

Table 6.3. *Information about Company A (received from Company A's representative)*

Total turnover $	2187000000
EBIT $	16065000
Cash flow $	19575000
NOWC $	48600000
OTD (On-time delivery)	93,60 %
Personnel	1080
Total Inventory $	34425000
Finished goods	0
WIP	19305000
Raw Material	15120000

Table 6.4. *Information about a component manufactured by Company A (received from Company A's representative)*

Turnover $	46468627
Product pcs	87750
Inventory	9874583
Finished goods	0
WIP	5537511
Raw material	4337072

and local managers try to compensate for unreliability that has arisen from delays and variability in lead-time of global supply-demand networks, by over-ordering, double buffering, and applying competitive pressure on manufacturing and the central allocation organization (Christopher, 1998).

6.4 Example

Company A is a component manufacturer in an electrical utility manufacturing network. The factory employs over 1000 people and has customers around the world. It is an important unit, because it is the only supplier of the component for the entire network.

Basic information about the company and the component including inventory levels, lead-times and on-time-deliveries, is given in Tables 6.3 and 6.4 respectively. The total inventory is divided into finished goods, Work-In-Progress (WIP) and raw materials. Company A's manufacturing policy is assembly-to-order (ATO): when an order is received, semi-finished components are assembled and then the final product is delivered to customers. Consequently, WIP inventory is large. Finished goods are delivered to customers one or two times per week and customers own transportation inventories. Consequently, Company A does not have a finished goods inventory.

Company A works in close cooperation with suppliers. The cooperation is based on long term relationships and communication between Company A and its suppliers

Table 6.5. *Information about Company A's supplier (received from Company A's representative)*

Supplier V	Material 1		
Purchase volume $	4456222		
Purchase volume (pcs)	112050		
Supplier's OTD %	94%		
Lead-time / days	min	mean	max
Order handling time from downstream customer order to upstream component order	-	1.3	-
Supplier delivery time (including transportation time)	1	2	3
Supply principle (MRP, kanban, reorder point)	Reorder point		

Table 6.6. *Information about Company B (received from Company A's representative)*

Market share %		34.80%	
Sales forecast 20XX		108000	
Sales $		18 684 667	
Product pcs		30 494	
OTD (On-time delivery)		95%	
Lead-time/days	min	mean	max
- Order lead-time EXW	6	14	35
- Production lead-time	5	12	20
- Transportation time		5	
Delivery time asked by customer EXW/ days			
Transportation mode		Truck x 1	

is frequent. Table 6.5 shows an example of Company A's supplier. As can be seen, the supplier's on time delivery is very high and the lead-time is quite short because the supplier's plant and warehouse are located near Company A. The supplier's lead-times include transportation time and order handling time. Order handling times can vary and contain non-productive time

Company B is an important customer of Company A and its importance will increase in the future. Company A reports that its on time delivery to Company B is 95% (Table 6.6). However, Company B reports that Company A' on-time-delivery is only 89.6% (Table 6.7). This might be caused by different measuring policies or embellishment of real data. Order lead-time without transportation is 14 working days. It contains order handling time, production time, and order backlog (1.3 days + 12 days). Production time and order handling time might also contain a backlog. In accurate supply-demand network analyses the consistency of lead-times has to be investigated very carefully because, for example, production time might contain unnecessary waiting times. It is important to eliminate these waiting times.

Based on the information about Company A, its suppliers and customers, ASDN software calculates different kind of measures (Figure 6.3). The suppliers' on time deliveries (OTD) vary between 92% and 95%. These numbers are acceptable so there are not many problems in raw material deliveries. The biggest problem is the

Table 6.7. *Information about Company A (received from Company B's representative).*

Purchase volume $		18 684 667	
Purchase volume (pcs)		30494	
Supplier's OTD %		89.6%	
Lead-time / days	min	mean	max
Order handling time from downstream customer order to upstream component order	1	3	5
Supplier delivery time (including transportation time)	2	7.4	9
Supply principle (MRP, kanban, reorder point)		MRP	

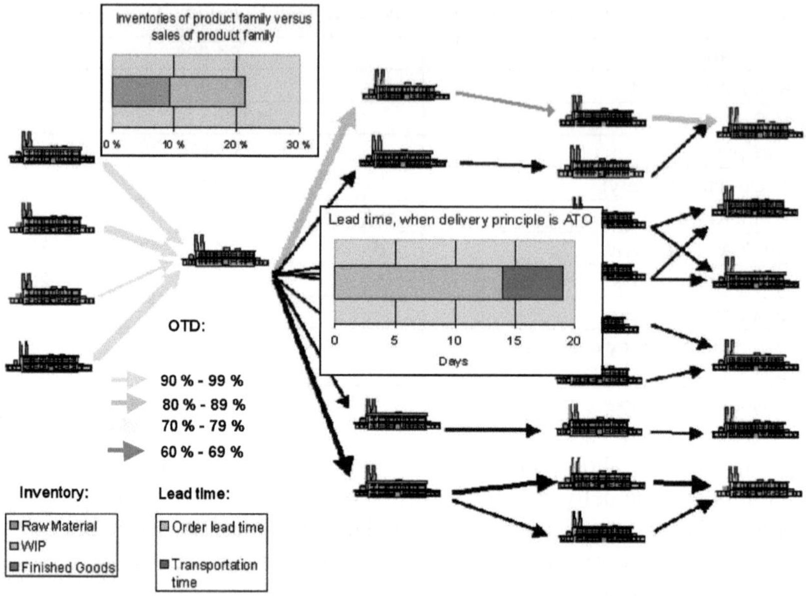

Figure 6.3. *The supply-demand network architecture*

inventory of the product family. It is 21.25% from the sales of product family, which is very high compared to the inventory of Company A (15.74%). The inventory turnover of the product family is 78 days. This is extremely high and causes problems in meeting changes in customer demand. High inventory levels tie up large amount of working capital. In this company, the biggest inventory is the WIP inventory so there is more cash-flow tied up compared to the situation when the inventory consist of raw material. A high WIP level might be a result of bottlenecks. If the inventory consisted of raw material inventory there would be the possibility to make changes to production plans and manufacture other products, but when the inventory consists of WIP such changes are not possible. These results help to understand why the inventory is so high and if there is a possibility to reduce inventories without decreasing OTD

ASDN software can be used to collect information about financial and material flows as well as data related to reliability of the supply-demand network. Then, by using modeling functionality, the user can build different development scenarios and see how the entire network is affected by changes in procedures and nodes. ASDN software can also be used as a visualization tool to describe problem parts of the chain and demonstrate the financial potential of improvements.

6.5 Discussions and Conclusions

The ASDN software presents an attempt to develop a coherent framework for decision-making about the supply-demand network structure. It has been assumed that the supply-demand network structure should follow the supply chain strategy.

Postponement/speculation and leanness/agility supply chain strategies can be described. Their connection to the supply-demand network structure decision-making that includes decisions on the number and location of companies in each tier and choices of transportation channels can be analysed by comparing alternative network configurations.

Structural decisions deal with issues such as the number and location of companies in each tier and choices of transportation channels. Ernst and Kamrad (2000) divide different supply chain structures according to the combined levels of modularization and postponement. The framework includes four types of supply chain structures: rigid, postponed, modularized, and flexible (Figure 6.4). The rigid structure represents the traditional, vertically integrated supply chain. With a rigid supply chain structure, the objective is to gain competitive advantage through economies of scale and high availability by maintaining large inventories of finished goods. The methods are similar to those illustrated in the speculation strategy. The flexible supply chain structure represents the other extreme with a greater number of suppliers and distributors. With postponement and modularization, the means to achieve high service level are in parallel with the postponement and hybrid strategies. Besides the two extreme structures, the framework includes two intermediate structures: postponed, with a high level of postponement but a low level of modularization, and modularized, with a low level of postponement but a high level of modularization.

To the best of our knowledge, there are no explicit descriptions of supply chain structures that correspond to leanness or agile strategies. However the same logic (structure follows strategy) should be followed when devising them. In agile production, excessive capacity, which is available on very short notice, is reserved. As the main goal of leanness is elimination of all waste, it can be suggested that the number of companies/facilities is less than when agility strategy is pursued.

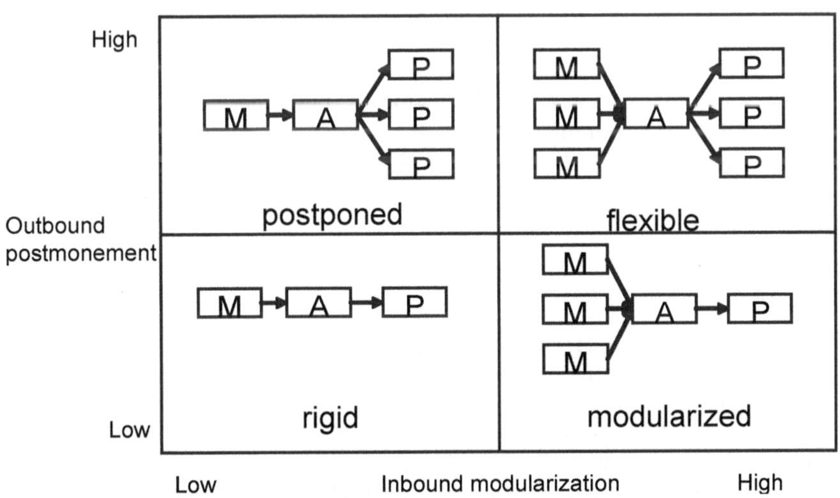

Figure 6.4. *Framework for the supply chain structure (M=manufacturing, A=assembly, P=packing) (Source: Ernst and Kamrad 2000)*

Table 6.7. *Supply chain structure characteristics of leanness and agility strategies*

	Leanness	**Agility**
Number of facilities in each tier	Low	High
Location	Where labour costs are low	Close to customer
Transportation channels	High capacity, low speed	Low capacity, high speed
Order decoupling point	Make-to-order, assembly-to-order	Make-to-stock, assembly-to-order

Furthermore, as the leanness strategy is the most suitable for high production volumes, labor costs play a significant role when operating in global markets. This is a basis for making decisions on facilities' location. The production volume also affects the choice of transportation channel, together with the intended level of responsiveness (Table 6.7).

This work on tool development is rather schematic and by no means complete. Improving agility or leanness may include industry specific issues (Helo, 2004), which need to be taken into account when conducting network analysis. It also follows that customers' needs can be satisfied with different types of strategies. For example, both leanness and full speculation strategies are suitable on markets with stable demand. So it is questionable whether similar supply-demand network structures can facilitate fulfilment of both. The ASDN software works well comparing this type of potential scenarios. Furthermore, it is necessary to mention that the approaches described implicitly take a static perspective on supply-demand networks. This perspective presupposes that a set of trading partners does not change much. The so called "chaotic" networks and "resilient supply chains" are an emerging trend in supply chain management. This phenomenon as well as the above mentioned issues should be taken into consideration when conducting applications in the field of supply-demand design and modeling.

Acknowledgements

The authors would like to acknowledge the EGLO programme, the Ministry of Transport and Telecommunication and the ABB Corporate Research Center for funding the research. ASDN Software is available for downloading at http://asdn.sourceforge.net.

References

Christopher, M. and Towill, D. (2001) "An integrated model for the design of agile supply chains", International Journal of Physical Distribution and Logistics Management, Vol. 31, No. 4, pp. 235–246.

Christopher, M. (1998) Logistics and Supply Chain Management: Strategies for Reducing Cost and Improving Service. London: Pearson Education.

Ernst, R. and Kamrad, B. (2000) "Evaluation of supply chain structure through modularization and postponement", European Journal of Operational Research, Vol. 124, No. 3, pp. 495–510.

Gattorna, J.L. and Walters, D.W. (1996) Managing the Supply Chain: A Strategic Perspective. London: Macmillan.

Hadley, S. (2004) "Making the Supply Chain Management Business Case", Strategic Finance, Vol. 85, No. 10, pp. 28–34.

Harland, C.M., Lamming, R.C., Zheng, J. and Johnsen, T.E. (2001) "A taxonomy of supply networks", Journal of Supply Chain Management, Vol. 37, No. 4, pp. 21–27.

Helo, P. (2004) "Managing agility and productivity in the electronics industry", Industrial Management and Data Systems, Vol. 104, No. 7, pp. 567–577.

Helo, P. (2006) "Agile production management: An analysis of capacity decisions and order-fulfilment time", International Journal of Agile Management Systems, Vol. 1, No. 1, pp. 2–10.

Helo, P., Xiao, Y. and Jiao, R.J. (2006) "A web-based logistics management system for agile supply demand network design", Journal of Manufacturing Technology, Vol. 17, No. 8, pp. 1058–1077.

Hicks, D.A. (1999) "The state of supply chain strategy", IIE Solutions, Vol. 31, No. 8, pp 24–29.

Hines, P., Lamming, R., Jones, D., Cousins, P. and Rich, N. (2000) Value Stream Management: Strategy and Excellence in the Supply Chain. London: Pearson Education.

Karmarkar U. (1993). Manufacturing lead times, order release and capacity loading. In: Handbooks in Operations Research and Management Science, Logistics of Production and Inventory 4:6. Eds. A.R. Kan Graves and P. Zipkin, Amsterdam: North-Holland.

Graves, S.C., Rinnooy Kan, A.H.G. and Zipkin, P.H (1993) Handbooks in Operation Research and Management Science: Logistics of Production and Inventory. Amsterdam: Elsevier Science Publishers.

Naylor, J.B., Naim, M.N. and Berry, D. (1999) "Leagility: Integrating the lean and agile manufacturing paradigms in the total supply chain", International Journal of Production Economics, Vol. 62, No. 1-2, pp. 107–118.

Pagh, J.D. and Cooper, M.C. (1998) "Supply chain postponement and speculation strategies: How to choose the right strategy", Journal of Business Logistics, Vol.19, No.2, pp. 13–33.

Richards, C.W. (1996) "Agile manufacturing: beyond lean?", Production and Inventory Management Journal, Vol. 37, No. 2, pp. 60–64.

Schary, P.B. and Skjøtt-Larsen, T. (1995) Managing the Global Supply Chain. Denmark: Handelshøjskolens Forlag.

Supply Chain Council (2008). Supply-Chain Operations Reference - SCOR Model 8.0. Available at: www.supply-chain.org

Vonderembse, M.A., Uppal, M., Huang, S. and Dismukes, J.P. (2006) "Designing supply chains: Towards theory development", International Journal of Production Economics, Vol. 100, No. 2, pp 223–238.

Womack, J.P. and Jones, D.T. (1996) Lean Thinking: Banish Waste and Create Wealth in Your Corporation, New York: Simon and Schuster.

7 Framework for Developing an Agile Future-Proof Supply Chain

Hossein Sharifi
Hossam Ismail
Iain Reid
The University of Liverpool

Abstract Two of the key elements that define a supply chain are "product", as the output of the business, and "supply chain operations", as the means of delivering the output. The processes of designing and developing each of these two elements are highly inter-related across more than one dimension. Many of the drawbacks in the success and sustainability of supply chains often relate to the misalignment of these two elements in one or more dimensions. In this chapter an integrated approach is proposed to facilitate the dynamic and simultaneous design and development of products and supply chains, thus contributing to the notion of agile supply chains. A framework is developed and, through a field case study observation, a number of issues raised in the framework are discussed and validated. An implementation model is also proposed in which the practical aspects of the framework stages are presented.

Keywords Agile supply chain, Design for supply chain, Supply chain design

7.1 Introduction

The concept of the agile supply chain is advocated as a new way forward for business networks to succeed in the highly changing and turbulent business environments. The main focus is in running businesses in network structures with an adequate level of agility to effectively respond to changes, as well as proactively anticipate changes and seek emerging opportunities. Agile supply chains are, therefore, those with the ability to rapidly align their structure and operations to the dynamic and turbulent requirements of the demand network. This paper proposes that key factors relating to how an agile supply chain can be developed, implemented and improved through the merger of two main processes: *Supply Chain Design* (SCD) and *Design for Supply Chain* (DfSC). The former is concerned with determining the network's strategy, designing its structure, processes and operations, and aligning its main constituents. The latter, which in practice is viewed as part of the new product development process, is concerned with designing the product while taking into account the impact on the performance and success of the supply chain. Designing a product for the supply chain results in both product improvements as well as enhancing the ability of the supply chain to operate effectively. In the literature each of these areas has been associated with the capabilities and characteristics required for achieving agility. The idea proposed and examined in this research is that a balanced approach to the two aspects of "supply chain design" and "product development" provides the required ground for developing agility in demand networks. This chapter presents the development of a new approach for the simultaneous product/supply chain design. A structured conceptual framework is derived, which addresses both the strategic and operational level of the design of an agile supply chain. The idea and the conceptual model are first developed via a review and analysis of the literature and is subsequently tested via a case study approach. Based on the case study material a framework is developed linking the key factors connecting market, product features, company capabilities and supply chain integration. The framework is further detailed by identifying a preliminary practical implementation model.

7.1.1 Agility, Supply Chain and Agile Supply Chains

The 1990s are associated with two important considerations in the area of business and operations management. First were the concerns around the high level of change and uncertainty in the business environment, which led to the concept of agile systems as presented by Nagel and Dove (1993), Kidd (1994), and more recently by Sharifi and Zhang (1999), Zhang and Sharifi (2007) and Ismail *et al.* (2006). Second of these concerns were the fundamental changes occurring in the principles of competition which have attracted academic and business interest. The latter resulted in a shift in thinking by viewing the supply chains as units of competition. Work by (Goldman *et al.*, 1995; Bowersox *et al.*, 1998; Christopher, 1998) are indicative of this change in thinking. Extensive work has been carried out in the area of supply chain management (SCM) from which novel approaches have been developed (Fine, 1998; Cox, 2000; Kehoe *et al.*, 2007; Sharifi *et al.*, 2002) to identify and understand the basic constituting elements of demand networks and their DNA.

The concept of agile supply chains was introduced (Harrison *et al.*, 1999) to transfer and apply the winning strategy of agility to that of supply chains addressing these as the newly accepted units of business. The idea of agility in the context of supply chain management focuses on "responsiveness" (Lee and Lau, 1999; Christopher and Towill, 2000). The drivers behind the need for agility in supply chains are similar to those that drove the introduction of the agile manufacturing concept and stem from the rate of change and uncertainties in the business environment. The operational dynamics of the extended supply chains contribute further to the uncertainties in the business environment and hence the vulnerability of the supply chain to change (Svensson, 2000). This situation has led to concerns over the slow growth of integrated supply chains (Lummus and Vokurka, 1999).

A classic definition of supply chains or demand networks is that they are entities developed from company collaborations formed to fulfil a business objective by delivering value to customers and the supply network companies by appropriation. These networks can be created based on either a predetermined design and plan, or emerge as the result of spontaneous needs in the course of planning and design of operations.

In the existing frameworks for introducing agility in supply chains such as those proposed by van Hoek *et al.* (2001), Christopher (1998, 2000) and Harrison *et al.*

(1999) mainly the same logic is applied to the original concept of agile systems and manufacturing. These, mainly, consist of a holistic view of the subject leading to practices and application of already proven concepts such as lean thinking, decoupling and postponement. Christopher (2000) suggests a three-level framework bringing together the various strands which contribute to developing the agile enterprise. These include rapid replenishment and postponed fulfilment, individual approaches such as lean production, organisational agility, and quick response, and finally any specific detailed actions needed to be undertaken.

The design of production and distribution systems has been an active area of research over the last 30 years (Van der Vorst and Beulens, 2002). Attention has also focused on the performance, design and analysis of supply chains as a whole (Beamon, 1998). The literature on the subject has focused mainly on the operational level and the physical structure of supply chains. However, growing attention has been specifically placed on the strategic issues related to the design of supply chain. Work by researchers such as (Fisher, 1997; Lamming *et al.*, 1999; Kehoe *et al.*, 2002) have contributed to the areas of determining strategic direction, formation and alignment models and implementing methodologies for demand networks. Approaching development and management of demand networks through alignment of strategies and operations within the networks has been a focal point in many recent works (Kehoe *et al.*, 2007; Henderson and Venkatraman, 1993; Lufinan *et al.*, 1993). Fine (1998) warns that when firms do not explicitly acknowledge and manage supply chain design and engineering concurrently with product and process design, engineering implementation problems often follow. To address this problem, a three-dimensional concurrent engineering (3-DCE) approach, where the simultaneous and coordinated design of products, manufacturing processes and supply chains are carried out, has been proposed by Fine (1998).

The importance of coordinating the development of product design and manufacturing process design is well recognised and various concepts such as "design for assembly, manufacture and operability, etc." have achieved a considerable impact on manufacturing (Huang, 1996). In addition to concurrent engineering, there has also been an increasing, and somewhat parallel, emphasis on synchronizing supply chain management decisions with product design decisions (Hult and Swan, 2003; Joglekar and Rosenthal, 2003; Lee and Sasser, 1995).

7.2 Framework for Agile Supply Chains: A Balanced Approach

Taking the basic principle of agility combined with the formal definition of supply chain agility into account, it can be suggested that for demand networks to be competitive they should achieve a sufficient level of agility that corresponds to the level of change and uncertainty in the overall as well as individual business environment. According to Sharifi *et al.* (2006), two main dimensions of physical/information and power/relationships/behaviour within the context of SCM are pivotal in determining supply chain strategy with regard to agility of the chain.

Developing a strategy for supply chains to become agile encompasses two issues. The first issue is how to identify the main building blocks for determining, developing and maintaining agility in a supply chain. The second issue is to determine how the model should be interpreted in terms of action and implementation steps. The second issue refers to the proposal of practical views which should incorporate the steps required to study and analyse the network, determine the strategies to approach agility within the network, and the means via which the agility can be introduced. A methodology for this purpose is proposed by Ismail and Sharifi (2005) and Ismail *et al.* (2006).

The first issue forms the main subject of this chapter. What elements contribute to the move towards agility in a supply chain? In simple terms a supply chain incorporates two main elements: product and supply chain/network. A supply chain is formed and managed, and subsequently decomposed or restructured corresponding to specific needs or emerging opportunities in the business environment. The design of a network of business entities is, therefore, preceded by specifying a fundamental business objective. This objective invariably involves developing, producing and delivering a product/service to customers (end users) for a financial reward. This product or service is subject to a design and development process comparable to that of the supply chain design. The two elements of product design and supply chain design have already been the focus of many researches in this area (Fine, 1998). There are, however, many remaining questions in this area. For example, what are the elements of these two processes, how do they interact and interrelate; how to facilitate the coordination of these two processes; and how a concurrent approach to them can enhance the supply chain responsiveness and agility. This research proposes a balanced approach to the implementation of these two processes and provides a practical vision on how supply chains' competitiveness can be sustained and further improved using this approach as shown in Figure 7.1.

The integration of these two viewpoints is influenced by a number of key internal and external factors that affect the supply chain strategy, as well as how the proposed approach can be formulated and applied. The key factors are numerous but generally can be grouped as follows:

- **Market and business environment factors**: Market factors cover aspects relating to the size of market, level of competition as well as type of market/industry sector, amongst others. They also take into account the stage in the product life cycle that the market is currently operating at as well as the rate of new product introduction. From a customer point of view, they consider the level of customer involvement in specifying product specification/features (requirements) as well as the position of the company in the supply chain with respect to the end user. The business environment factors cover aspects that include legislative, economic, social and environmental factors that impact on the company's ability to achieve its intended strategy.
- **Product factors**: These include product complexity and level of technology and innovation involved in developing and manufacturing the product. Also to be considered are the level of services involved in supporting the product from production through distribution to after-sales support. The complexity of product features is dependent on the level of certainty with respect to those qualifiers and order winning factors that differentiate the product.

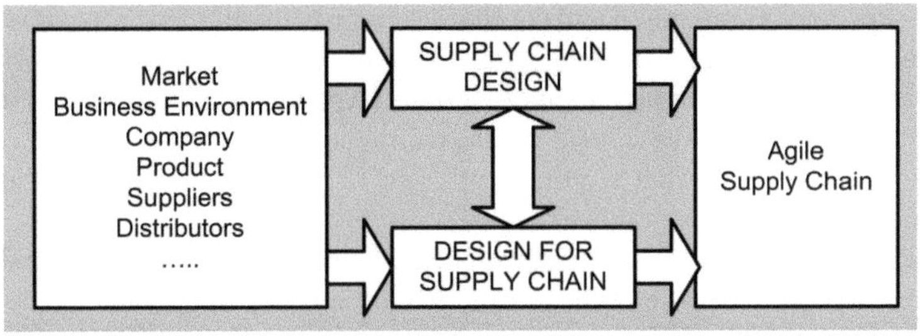

Figure 7.1. *Agility in supply chains through SCD and DfSC*

- **Company**: Company factors are predominantly concerned with internal capabilities. These range from the ability to understand the dynamic nature and requirements of markets through to efficiently and effectively satisfying these requirements. It also considers the company's ability to identify a strategy and rapidly roll out internal and external resources to meet the requirements for that strategy.
- **Supply chain**: Supply chain factors cover suppliers' capabilities and availability. They also cover how the chain operates, the speed and level of effort required to set up, align and maintain. It also addresses the necessary nature and level of communication, trust, and balance of power in the supply chain. The supply chain's responsiveness and resilience to changes both within the supply chain and in the business environment are also critical.

7.3 Supply Chain Design

As a unit composed of various parties with overlapping and conflicting interests supply chains are composed and subsequently decomposed or restructured depending on the specific needs or emerging circumstances in the business environment. To be successful, the process of assembling the supply chain must be carried out efficiently. The emergence of such systems should be based on a specific and detailed design process in which the characteristics of the chain or network are identified and implemented.

The basic role of SCD is to provide an optimal platform for efficient and effective SCM and act as a bridge to connect supply chain strategy and the supply chain operations. SCD can be considered from two view points, strategic and operational. From a strategic point of view, SCD refers to the process of determining all required components of the supply chain including its structure and operations aligned to customer requirements and supply chain strategy. This viewpoint addresses a wide range of strategic and tactical infrastructure issues that are specific for each enterprise (Harrison, 2001). SCD can also be referred to as the process of devising the supply chain infrastructure and logistics elements which includes determining the location and capacity of plants, distribution centres, transportation modes, fleet and lanes, production processes, logistics information exchange patterns, etc. According to Appelqvist (2004), SCD covers two main dimensions including pre-determining and reengineering, and optimization and continuous improvement. These two dimensions follow an analogous process that starts from design requirements analysis and through to the set up of supply chain objectives (Appelqvist, 2004). From a review of the literature, SCD can be considered as composed of five general stages:

i. Understanding of market requirements, and the current situation of the supply chain.
ii. Determining supply chain performance attributes based on an analysis of customer requirements and the current situation of the supply chain.
iii. Determining supply chain performance dimensions that stand for the areas where the supply chain attributes can be decomposed to more concrete performance dimensions.
iv. Translating supply chain dimensions into supply chain functions converting the conceptual supply chain to an actual supply chain.
v. Designing and examining all the components and aspects of desired supply chain against the market requirement and current situation. This is the most complex step and consequently costly and time consuming.

From a market point of view, the main purpose of constructing a supply chain is to meet some predetermined need in the market. Equally important, the supply chain is also there to support the continual growth of its members. The problem of how agility can be interpreted in terms of supply chain operational strategy and integrated within each of the network member's strategy is probably the main concern within the context of designing and managing an agile supply chain. Furthermore, issues such as value appropriation, power and relationships and incorporating these into the supply chain design are important to resolve. Once the design is underway, implementing a supply chain strategy with all its consequent operations and processes also needs to be addressed.

7.4 Design for the Supply Chain

The common approach to product development in a market driven business environment is to initially identify product features to best meet all or most market requirements. The supply chain network necessary for achieving these objectives is subsequently formed from those available resources (internal and external) that are capable of providing the planned specification and performance needs. During the development process, emerging resource and capability limitations are dealt with through forgoing certain market requirements or investing in new resources or searching for new external sources to meet those needs. Driven by the desire to maximise market potential, traditionally, the process starts with translating all market specified requirements into product features requirements, as shown in Figure 7.2(a). Product features are sometimes separated into qualifiers, winners and delighters (MacMillan and McGarth, 1996). However, this all encompassing approach often results in initial concept designs that are exceedingly ambitious.

Subsequently, the new product development (NPD) processes iterate through a number of internal stages whereby product features are often culled to achieve a viable product. This is a costly and time consuming process that entails wasted effort by repeatedly redesigning and redefining products features. To partially address this, new practices in the area of product development have emerged. Some of these approaches are at the management level such as "Concurrent Engineering" while others are at the detailed level such as "Design for Manufacture and Assembly" and "Design for X". Similarly, other approaches such as "Postponement" have emerged to resolve issues arising from operating in a "mass customised market" with varying lead-times. However, these practices, whilst beneficial, have not been convincingly applied to resolve difficulties involved in product development process across networks. These problems include addressing network limitations, fairness of value appropriation and cost distribution, burden of responsiveness, changing market circumstances and demand dynamics and integration of the entire process, and so on. Nevertheless, the adoption of new internal approaches has proved very valuable and as a result has changed the way that managing the design and development process of products is viewed.

An alternative approach to improving the process is based on "Design for the Supply Chain" and starts the NPD process from an achievable point with respect to product features as shown in Figure 7.2(b). These features represent those that the existing supply chain network can deliver rapidly if required. Guided by the full market-specified feature list, these initial capability features are extended as a result of further collaboration with suppliers and extending the supplier range. The advantage of this approach is that the product is viable at any stage of the product design process. The product development process is maintained until time and cost constraints dictated by projected investment returns are reached. This approach enables the supply chain to respond quickly to emerging opportunities and, furthermore, facilitates the introduction of practices such as using common product platforms, modularity, product/component reuse and design outsourcing.

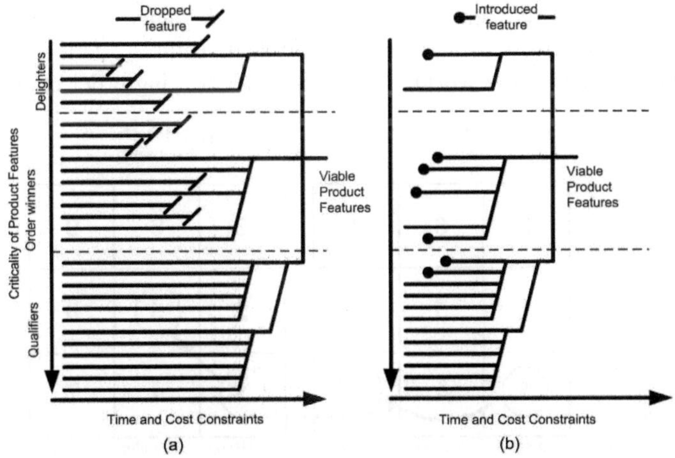

Figure 7.2. *Product feature development approach: (a) traditional, (b) Design for Supply Chain*

7.4.1 Growth Strategy

The above process changes somewhat depending on product newness and the level of market pull or technology push involved. The process is also mitigated by time issues such as speed to market and product introduction clockspeed.

Using the extended Ansoff matrix as a point of reference (Figure 7.3) there are a number of transitions a company can undergo from an existing market position (1):

i. Companies traditionally extended the sales of their existing products by moving from sector 1 to sectors 2 and 3 through cost and operational efficiencies and where possible align their existing supply chain to meet this new shift in emphasis.
ii. Extending the product range through a shift from sector 1 to sectors 4, 5 and 6 involves a redesign or modularisation of the product to capitalise on new opportunities in customisation and product platforms (Ismail *et al.*, 2007). A redesign of the supply chain is often required with a shift in emphasis from cost to flexibility.
iii. A new product introduction strategy, represented by a shift from sector 1 to sectors 7, 8 and 9, is the most risky but offers the company the opportunity to fundamentally redesign the supply chain to meet the new product needs. However, in this case, it is critical to identify at an early stage the subsequent growth strategy of the proposed new product. For example, a shift from sector 1 to 7 will involve partnering with innovative suppliers. However, if the subsequent strategy is to move to sector 5 then it is important that selected suppliers are also capable of such flexibility.

7.4.2 Product Life Cycle and Clockspeed

Other industry sector factors that impact on the approach are the present market position in the product life cycle and product introduction clockspeed. A product entering the market in the "introduction" and "growth" stages, as shown in Figure 7.4, has a degree of freedom in specifying product features. This is more so for products that are new-to-the-world. In this case, identifying order qualifiers, order

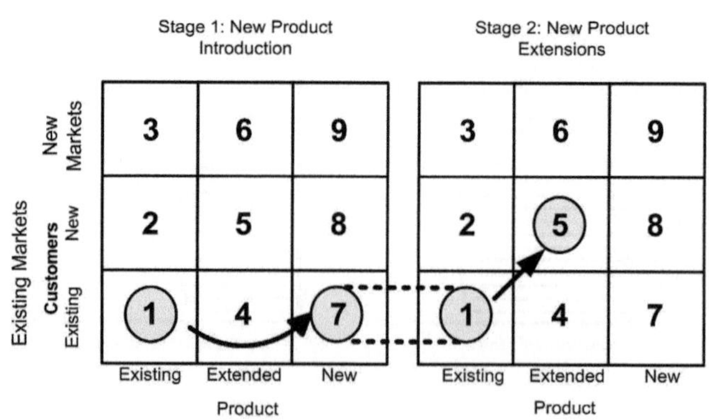

Figure 7.3. *Extended Ansoff matrix for growth strategy*

winners and delighters is not firmly established. As the market matures, newly introduced products will have to conform to a minimum set of requirements, and, therefore, the selection of product features is more constrained. Products launched at the decline stage often try and capitalise on cost differentiation through economies of scale and (i.e. where features may be withdrawn or downsized to re-stimulate demand for a basic product).

From a clockspeed point of view, markets that provide only a small window of opportunity, due to high product introduction clockspeed, create a barrier for incoming new products to establish a market position. Quite often the only possible approach is to leap-frog and provide a step change in product features. This requires a highly innovative supply chain strategy.

The adoption of an approach based on "Design for Supply Chain" has to be defined and managed strategically. The downside of little or no strategic planning is the possible suppression of ideas that do not fall within the supply chain capabilities. The approach should also go beyond the limited financial assessment of "make or buy".

7.5 Integrated Framework of Simultaneous SCD and DfSC

An integrated framework is developed from the above discussions to provide a practical approach to the simultaneous development of the two elements. First it is possible to recognise how an approach can be introduced in the management of supply chains that provides practical means for designing the products and processes "for" the supply chain as a unit of business and competition. The idea of integrated SCD and DfSC, however, needs to be supported by means for understanding of the industry requirements as well as interpretation, analysis and implementation. A number of recent studies (Lee and Sasser, 1995; Forza *et al.*, 2004; Fixson, 2005; Petersen *et al.*, 2004) in the area of integrating the design and supply chain processes have emerged. These range from focusing on a single attribute (e.g. lead-time) to extending product ranges through replaceability and common interfaces.

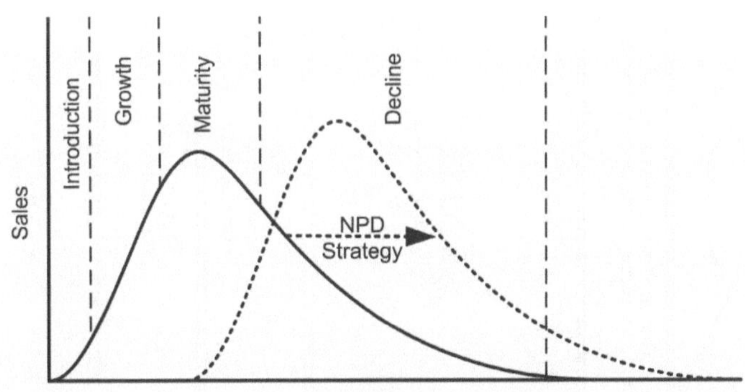

Figure 7.4. *New product lifecycle*

In this chapter an agile supply chain development framework is proposed, shown in Figure 7.5. The framework derives its structure from that of the principles of Quality Function Deployment (QFD) (Hauser and Clausing, 1988) and is driven by market needs or the voice of the customer where these are translated to product features. However, it includes a number of key stages involving the alignment of features to the basic strategic and operational supply chain properties. The framework elements can be summarised as follows:

i. Feature extraction and classification, where product features are identified and grouped based on their criticality into order qualifiers, order winners and delighters. A cross impact analysis of these in terms of interdependency and conflict is carried to further prioritise these features.
ii. Feature assessment, where features are assessed in terms of how they are aligned to one or more possible strategic product differentiators. These differentiators cover cost, quality and delivery and the extended properties of flexibility, robustness, innovativeness and service. These properties are derived from Miltenburg's (1995) approach to defining manufacturing strategy and operational requirements.
iii. Business environment assessment, which addresses all non-product feature based factors that could impact on the current and future potential of the product.
iv. Company capability assessment, involves matching the product features to company capabilities with the aim of constructing a company view of the ideal product. At this stage, features are also assessed along the line of "make or buy" (see, for example, Dekkers, 2000, and Dekkers and van Luttervelt, 2006).

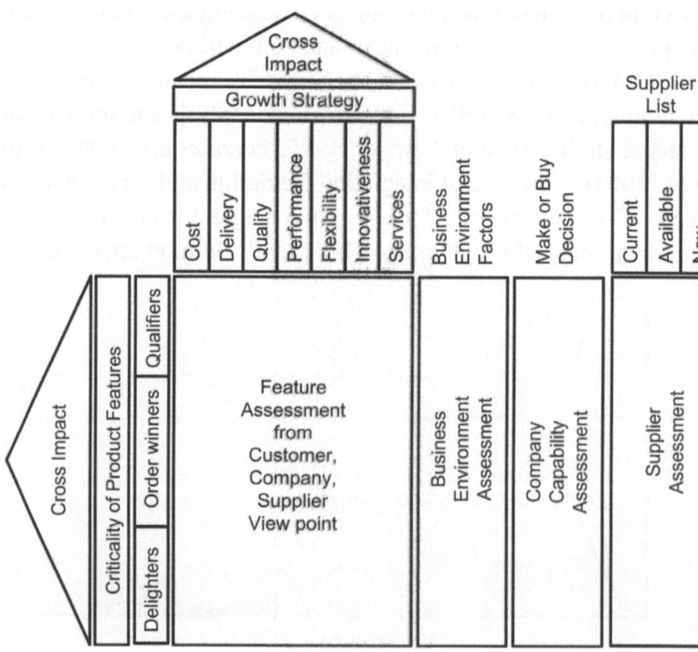

Figure 7.5. *Agile supply chain development framework*

v. Supply chain assessment, involves assessing the existing and potential supply members across the product feature requirements and building an ideal supplier profile for each.
vi. Feature clustering and alignment is subsequently carried out in terms of supply chain capabilities to ascertain what can be achieved immediately if time is critical and what is possible to achieve if cost is not a constraint.

The result from the above framework is a selection of those necessary product features that can rapidly meet market demand on one hand and the potential company growth strategy on the other. The framework identifies supplier profiles and matches these to existing and potential suppliers. Its strength is derived from the ability to integrate a market, product, company and supply chain points of view under one assessment framework. The resulting product is therefore a compromise that fulfils market needs from one end and supply chain agility capability at the other end. The application of the framework does not exclude the use of well established tools at each of the stages of market research, development, outsourcing, manufacture and distribution but sets a common platform for linking these tools.

7.6 Preliminary Validation of the Framework

The concepts and the integrated framework were examined by adopting a case study research methodology to explore and validate the concepts and approaches proposed for developing agile supply chain. This approach fits well within the case study research category as it is recognised as being particularly valuable for examining "how" and "why" questions (Yin, 1994). Voss *et al.* (2002) have also recommended this approach for theory testing, but more importantly for theory development. Considering the dimensions of the proposed model the multiple case study method (Yin, 1994) was chosen. The prime method of data collection included semi-structured interviews combined with sector specific sources of data. The focus of the study and the results presented below was intended to validate the conceptual framework structure as well as to assess participating companies' perception of the framework. A sample of four companies was selected as a basis for the research, the results of which are reported below.

7.6.1 Case Study Companies

In this study, four SMEs (Small and Medium Enterprises) were examined. The cases were chosen from a group of companies with whom the research group had existing research and support connections. The following criteria were considered in the selection of sample companies represented in Table 7.1:

i. The company is an OEM (Original Equipment Manufacturer) with a good track record and position in the market.
ii. A reasonable size of supply chain is connected with the company.

iii. The company's products are at a minimum level of complexity in terms of market requirements, features, technology, and process so that the in-house design activity plays a relatively important role in their development.

7.6.2 Case Study Design and Results

A semi-structured interview was used for the study. A questionnaire consisting of four main parts was developed to cover company profile, issues related to the product design, strategies and, structure and operations of the supply chain. The last part of the interview was designed to capture issues relating to inter-relationships and cross impacts of product design and supply chain issues on each other and on total competitiveness of the companies. The technical directors of the companies were interviewed. Prior to the interview a copy of the questionnaire was emailed to the company for the purposes of familiarisation.

a) Products, Features and Design

Various aspects of the companies' products and strategies, future growth objectives and span of activities in developing new products were examined. As shown in Table 7.1 all cases produce highly customised products to a relatively competitive market. Furthermore, most of the company products are at a high level of complexity in relation to the company size. Each company was asked to select one of its core products and to specify its drivers and to provide a breakdown of the product in terms of internal and external product features. These features were then grouped under the three categories of order winners, order qualifiers and delighters as well as against how they contributed to strategic priorities of the company such as Cost (C), Delivery (D), Quality (Q), Performance (P), Flexibility (F), Service (S) and Market (M). As presented in Table 7.2 products are mainly technology and design driven with 3 to 8 main features, which in most cases are qualifiers. Determination and

Table 7.1. *Case companies' general information*

Co.	Sector/ Category	T.O M£	Product range	No of Prods	Product type
C1	Manufacturing	£6 m	Sport and play units and items	32	Customised, modular, packed to order
C2	Manufacturing	£600K	Eyebath and shower (for industrial application), combination	17	Customised, modular, engineered to order, make to order
C3	Software	£300K	Information Kiosk, Software	3	Engineered to order, customised software
C4	Manufacturing	£850K	Ultrasonic cleaning system	4	Customised, modular, make to order

choice of product features are mainly driven by the company with some influence from customers and competition.

In all cases companies have major plans for growth in both dimensions of market and product. However, despite the fact that most of the case companies apply the usual steps in new product development and introduction, in almost none of the companies considered was the design of the product linked to the supply chain design process.

Table 7.2. *Products and processes*

	Product Characteristics			**Features in product**		**Qualifiers/Order Winners/Delighters**	**Product drivers & Priorities** Cost (C), Delivery (D), Quality (Q), Performance(P), Flexibility (F), Service (S), Market (M) [in the order of importance]	**Design Intensity**	**Growth strategy/ Current and target position in ANSOFF matrix**	**Usual stages in NPD** Marketing = M Concept&Design=CD Detailed/Design= DD Prototype = PP Test = T
	Trend	**Market perception**	**Complexity ASPECTS (LEVEL)**	**Internal**	**External**					
1	Seasonal (Mar to Jul/Aug)	Customised Leisure	Varieties, no. of components (Medium)	1	7	Q= 6 OW/D = 2	Company, Competitor P->(Q&F&S)-> (D&M)	L	Existing products, extending **From 1 to 5.**	M, CD, DD, PP, T
2	Increase	Innovative & Customer Driven	Varieties, no of comp (Low)	2	3		Customer Technology Driven P->(Q&F&S)-> (D&M)	M	**From 1 to 5/6.**	M, R&D, CD, DD, T, PP
			Process, variety, technology (Low)	3	0	Q= 6 OW/D = 1	Partnership Agreement Sources 4 -7 products	L	Existing and extending product, new M, C, & P ANS	N/A
3	Increase from Jan.-March	Cutting-edge	Process Hardware (Medium) Software (High)	1	6	Q= 6 OW/D = 1	Company, Technology, Competitor, Customer (Q&P)->C->F-> S->(D&M)	M	Extending product, new market and customer **From 1 to 6.**	M, R&D, CD, DD
4	Increase	Techn. Driven	Tech. No of comp. Hardware (Medium)	6	2	Q= 7 OW/D = 0	Tech. driven, customer P->S->(Q&F)-> (C&M)->D	M	Existing products, new market &customer **From 1 to 9.**	M, R&D, CD, DD, T, PP

Table 7.3. *Supply chain of case companies*

	Suppliers			Relationship-type		Replaceability	Supply Chain		
	Total	Specialist	General	Market	Long Term/ Partner		% outsource	Supplier dependency	Problems with suppliers
C1	15	2	13	14	1	H, depdt on 1, 1 depdt upon	35%	H	H
C2	12	10	2	2	10	H	10%	L	M-L
C3	5	1	4	5	0	H	90%	H	H
C4	120	20	100	120	0	H	85%	H	L

b) Supply Chain Features

The studied companies' supply bases were relatively large in size though mainly combined of general/market and replaceable type. Companies were asked to identify their level of outsourcing in terms of those components or modules that require no further work other than assembly. This narrow definition resulted in excluding items that required cosmetic or further work for interfacing purposes only. With 10% to 90% level of outsourcing (average of more than 50%) the supply chain plays an important role in the case companies' business. The implication of this is a high level of company dependency on suppliers (3 cases out of 4), and a considerable level of problems in the supply chain operations and management (Table 7.3).

Another view of the companies' supply chains is shown in Table 7.4, which depicts the issues, priorities and approach of the companies in designing and developing their supply chains. The data shows in particular that despite actual involvement of the supply base in parts of product design (and hence development), the design of the supply chain (in terms of choosing partners according to the

Table 7.4. *Supply chain design criteria*

	Criteria in Supply Chain Design			
	Physical Issues	Capabilities	Operational	Involvement in Design
C1	Cost of transport, position	Expertise in painting, quality in tubing	Flexibility, availability	Powder Coating (Material Coating and Pigment)
C2	Locality.	Capable of distribution Standard components	Reliability	Laser Cutting of Components
C3	UK based	Response time	Availability	PC Technologies
C4	US Patent rights Epoxy Resin		Lead time, stock	Design Gearbox Module Chemistry & Chemical requirements

capabilities and operational criteria) is in no parts related to the products' design and development process.

c) Dimensions' Interrelationship

At this stage of the research the interviewees were made aware of the result of the above analysis, i.e. the segregation of the two processes, and two new questions were asked. First, respondents were asked whether they found the two processes, SCD and DfSC, interdependent and mutually exclusive, and if a simultaneous approach to these dimensions would contribute to the companies' ability to deal with the changes in the market and business environment, hence their agility. This resulted in positive answers with a strong level of agreement in all four cases. Next, the companies' specific issues with regard to the conflict between the two dimensions, from a supply chain management point of view, were examined. The areas considered in this section included supply chain problems in terms of operations, relationships and growth, and how they are connected to the design process. Table 7.5 presents a summary of the issues identified.

Table 7.5. *Problems in the supply chain and relation to the product design*

	Problems with SC			
	Operational	**Relationship**	**Growth Stage**	**Design Related**
C1	Delivery until wagon is filled up.	Locked by steel supply because of payment issues.	Standardization, variation in steel, modification on fixture.	Length of steel, Storage of steel, grade of steel.
C2	Oversized part is avoided by drawing imperative lines.	Contract with suppliers and buyers to secure fixed or stable price.	As the volume increases, machinery becomes a problem and suppliers' capability cannot catch up.	Laser cutting imposes redesign on some products.
C3	Tho balancing point between cost and quality, Increased price of components results in rearrangement of job carried out by supplier.	Locked in the relationship with metalwork supplier.	The company has become a key customer to the fabrication of the Kiosks.	No specific drawing and design parameter of the fabricated Kiosk. 19" Monitors dropped due to portability to a standard 15" resulting in 8 months redevelopment.
C4	Density heater does not fit the machine in the expected way. Gearbox supplier underestimate the motor of gearbox.	Lock-in with certain specific supplier.	Suppliers cannot meet the requirement demanded by new heater system.	No dual source for some components, Supplier is off all of August. Rely on sole distributor. No specifications dimensions, steep learning curve.

The findings of the case studies show that many supply chain problems could have been avoided if they had been considered during the product design stage. In most cases, late emerging problems could have been eliminated by redefining the product features and design, which in turn would have impacted upon the form and operation of the supply chain.

The above case studies provided valuable insight into the strong interdependence between two dimensions of supply chain design process (SCD) and the process of designing and developing products, and the contribution of an integrated approach with respect to agility. Segregation of product development processes and supply chain development and management has proven to undermine the advancement of capabilities necessary for responding and succeeding in environment characterised by turbulence.

7.7 Preliminary Implementation Model

The integrated framework proposed above is developed with the aim of providing a structured approach for future proofing a supply chain. In its current form, the framework does not prescribe any specific set of tools and in effect it was designed to be generic in structure independent of tool type. In the following section a number of tools to demonstrate the practical application of the front-end of the framework is presented and represented as shown in Figure 7.6.

The implementation model consists of a number of distinct stages as described in the following:

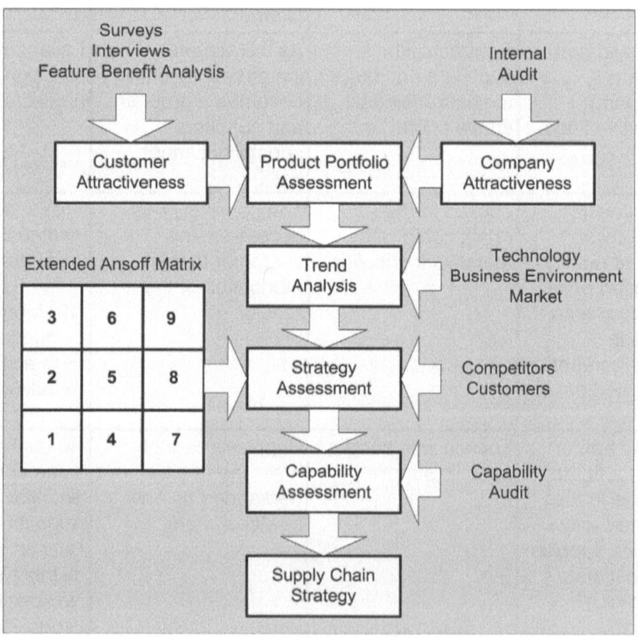

Figure 7.6. *Implementation model*

- **Product portfolio assessment**: This is where initially a company's product portfolio is assessed along two dimensions (Poolton *et al.*, 2006). The first dimension addresses the level of "customer/market attractiveness" for each of the products or product families. The assessment of product-customer attractiveness is carried out through internal review of the share of turnover as well as external assessment of customer feedback and perception. This assessment is extended to include identifying critical product features through tools such as interviews, surveys and more specific conjoint analysis.
  The second dimension is "company attractiveness" and covers the level at which the company's key resources are committed to developing and delivering each product. This dimension is an indication of the ease or otherwise by which the company develops and delivers the product and also includes an overview financial assessment of the contribution of each product.
- **Trend analysis**: This stage case covers an analysis of the current business environment, market and technology trends. This is carried internally through identifying the company's perception and externally through analysing trade association and governmental reports. An impact analysis is carried out for each of the trend factors. This covers amongst other factors:
  i. the company's level of control with respect to the identified trend;
  ii. the impact on the company in terms threats and opportunities;
  iii. the resources required to capitalise on the emerging trend;
  iv. the level of risk involved.
- **Strategy assessment**: The growth strategy assessment is based on evaluating possible strategies by reference to the extended Ansoff matrix. As discussed above, various growth strategies can be identified from the current position and proposed shift within the extended Ansoff matrix. In general terms, each position in the matrix is associated with a possible set of criteria from both a product differentiation point of view and accordingly a supply chain capability requirement.
  For product differentiation, the criteria used are those derived from Miltenburg's (1995) manufacturing strategy model. These criteria, in addition to the basic cost, quality and delivery, include innovativeness, flexibility, performance and service level. While general assumption can be derived concerning which of the differentiation criteria are more prominent in each of the Ansoff matrix cells it is not possible to universally fix these assumptions. Business environment changes and the specifics of each industry type will dictate which of the criteria would play a key role for growth. The criticality of these factors is derived from an in-depth understanding of the market. As an example, Table 7.6 demonstrates a view of how the differentiation factors are mapped and prioritised across the various matrix cells. These are listed in order of prominence for a specific analysis of a company involved in developing products in the construction industry.
  Each differentiating factor is further decomposed to identify those associated product specific features, evaluated against competitor's products, and

Table 7.6. *Example of an extended Ansoff matrix with differentiation factors*

		Existing (Product)	Extended (Product)	New (Product)
	New Markets	Price Delivery Flexibility	Performance Innovation Service Delivery	Innovativeness Performance Service Flexibility
Existing Market	New Customers	Price Delivery Flexibility Quality	Performance Innovativeness Delivery Flexibility	Innovativeness Performance Service Delivery
Existing Market	Existing Customers	Price Quality	Performance Price	Innovativeness Performance Service

accordingly clustered in terms of customer attractiveness along the lines of qualifiers, order winners and delighters.

- **Capability assessment**: Once a strategy and product features are identified the next step is to carry out an audit of the company's capabilities to ascertain the viability of the strategy from a practical point of view. The capability assessment is carried out across a number of factors covering product, process, people, operations and organisation with respect to the above critical factors. For each of the capability factors a set of measures are identified that address the requirements of the selected strategy and product features. A sample of such capability measures is given in Table 7.7.
- **Supply chain strategies**: A supply chain strategy is derived based on the company capability measures described above as well as the potential capability of the existing and proposed supply chain. Decisions on features with respect to "make", "buy" or "drop" are set at this stage based on a number of influencing factors as:

  i. criticality of each factor/product feature;
  ii. competition;
  iii. availability of resources (company and supply chain);
  iv. time constraints;
  v. cost constraints.

  At this stage product features are further classified in terms of availability and novelty to the company and world. They fall under a number of categories as follows:

  i. available and currently in existing products;
  ii. achievable with existing company resources;
  iii. new but obtainable from existing suppliers;
  iv. new but require engaging with new suppliers;
  v. new to the world.

For each of the features existing and potential suppliers are evaluated for the ability to align to current and future needs, and accordingly a further review of the strategy

Table 7.7. *Interrelationship of strategic priorities (outputs) and capabilities*

Strategic Priorities (Output)	Capabilities	Measurement Criteria
Delivery	• Availability of skilled labour to service an order • Technical knowledge of product to prevent glitches • Streamlined purchasing • Streamlined manufacturing • Appropriate levels of stock • Ample logistics capabilities	• Percentage on-time deliveries • Accuracy of inventory status • Average delay • Master Production Schedule performance/stability • Delivery time
Cost	• Cheap availability of raw materials through supplier networks • Appropriately skilled labour [not over-skilled or under-skilled] • Efficient purchasing • Lean manufacturing • Market reach [acquisition cost] • Ergonomic product design • Recyclable product components • Life Cycle Costs	• Unit product cost • Unit labour cost • Unit material cost • Total manufacturing overhead cost • Inventory turnover – W.I.P., raw material, finished goods • Capital productivity • Capacity/machine utilisation • Materials yield • Direct labour productivity
Quality	• Appropriate machinery • Correctly trained operators • Unambiguous definition of specifications • Manufacturing input conformity [supplier quality] • Manufacturing output consistency & conformity • Ability to satisfy market qualifying criteria	• Internal failure cost – scrap/rework, percentage defective/rejected • External failure cost – frequency of failure in field • Percentage unscheduled downtime reduction • Assembly line defects per 100 units • Percentage defect reduction • Percentage scrap value reduction • Percentage of inspection operations eliminated • Vendor quality
Performance	• Culture of innovativeness in the firm • Knowledge of competing products • Intimate knowledge of own product • Intimate knowledge of market requirements • Ability to convert customer requirements into design specifications	• Number of standard features • Number of advanced features • Number of additional value-added features [as compared to competitors] • Product resale price • Number of engineering changes • Mean time between failures
Flexibility	• Capable supply chain upstream and downstream • Ability [liquidity, credit rating, available suppliers] to purchase varied quantities of raw materials • Ability to efficiently manufacture varying quantities of existing products • Ability to service varied quantities of orders while maintaining service levels	• Minimum order size • Average production lot size • Length of frozen schedule • Average volume fluctuations that occur over a time period divided by the capacity limit • Number of parts processed by a group of machines • Ratio of the number of parts processed by a group of machines to the total number processed by that factory
Innovativeness	• Technical knowledge of industry • Capable machinery • Capable suppliers • Ability to predict future market requirements • Ability to satisfy these requirements today • Finding simple, ergonomic solutions to satisfy market requirements • Company-wide innovativeness • Agile manufacturing techniques	• Number of products in the product line • Number of available options • Number of engineering change orders per year • Number of new products introduced each year • Lead time to design new products • Level of R&D investment • Consistency of investment over time

Table 7.7. *(continued)*

Strategic Priorities (Output)	Capabilities	Measurement Criteria
Service	• Detailed customer information • Ample product knowledge and literature • Customer support team • After-sales maintenance teams • Customer Relationship Management capabilities	• Gap between consumer expectations and Management perception of those expectations • Gap between management perceptions of consumer expectations and the firm's service quality specifications • Gap between service quality specifications and actual service delivery • Gap between actual service delivery and external communications about the service

and feature selection is carried out. While the above model stages are presented in a sequential form, in practice, the stages are iterative rapidly converging on a suitable selection of features at one level and the corresponding supply chain configuration at another.

7.8 Conclusion

Future-proofing a supply chain depends on a number of factors. Key to these is an ability to operate in an agile and possibly opportunistic manner so as to be able to respond to market needs. An agility capability in supply chain companies in isolation is not sufficient as success is dependent on the effective integration of these companies. Supply chains need to reflect the requirements of the market and the business environment. Accordingly, flexible mechanisms are necessary to respond to the dynamics of the business environment. The chapter presented a conceptual framework, which addresses the issue of developing agile supply chains. It proposed an approach that integrates aspects relating to product development and supply chain development defined as "Design of Supply Chain" and "Design for the Supply Chain". For this to succeed, a calculated approach is required that takes into account the design of products with particular attention to the characteristics of the supply chain and it dynamics. The two aspects of SCD and DfSc, discussed in this chapter, interact with factors such as the market place dynamics, supply chain dynamics, business environment, technology, as well as with each other to support the dynamic characteristics of agile supply chains. The chapter validates the need for a framework via a case study approach in which four OEM companies were investigated. For example, in all case study companies the set of key suppliers that they are currently using is different from the set they initially started with at the introduction of the product. The reasons vary from those that are related to technical capabilities to those that come under operational and capacity issues, with the expected implication on costs and time. Case study companies were asked the question: "Knowing what you know now about the capabilities of your suppliers, would you have designed the product differently?". The answer was an unqualified "yes". The case study companies, despite being SMEs, have understood the proposed approach and appreciated the potential savings they could have achieved in previous projects. These results are characteristic of those found with respect to the introduction of Concurrent Engineering in the 1980s where the integration of the design process with manufacturing, assembly and modularity, etc. led to radical improvements in quality, cost, and flexibility.

The result of this preliminary analysis highlighted some of the issues that such a framework needs to address showing examples of avoidable supply chain problems. A number of models and methods are hence introduced to conceptualise the idea of a holistic approach to a future-proof agile supply chain. Further, an implementation approach with a view to developing practical solutions for agile supply chains is proposed. In particular an approach is proposed for strategy assessment in which the growth strategy is assessed by evaluating possible strategies using the extended Ansoff matrix. Alternative growth strategies can be identified from the current position and possible shift within the matrix where each position in the matrix is associated with a possible set of criteria which are chosen to be supply chain strategic priorities. This strategic orientation is also supported by the theory of "disruptive technology" and "sustaining technology" of Christensen (1997). He defines the move to capture either lower-end market through disrupting the competition via "cost/price" or satisfying non-consumption by offering them the opportunity, and/or moving towards up-market and hence poaching competitors' customers for which it means being "innovative" to supersede the competition by satisfying the more attractive customers. Through extending the proposed model and developing implementation tools it is expected to provide practical solutions for complex issues in supply chain management.

References

Appelqvist P., Lehtonen J.M., Kokkonen J. (2004),"Modelling in product and supply chain design: Literature survey and case study", Journal of Manufacturing Technology Management, Vol. 15, No. 7, pp. 675–686.

Beamon B.M. (1998), "Supply chain design and analysis: models and methods", International Journal of Production Economics, Vol. 55, No. 3, pp. 281–94.

Bowersox D.J., Closs D.J., Hall C.T. (1998), "Beyond ERP – the storm before the calm", Supply Chain Management Review, Vol. 1 No. 4, pp. 28–37.

Christensen C.M., (1997), "The Innovator's Dilemma", Harvard Business School Press, Boston.

Christopher M. (2000), "The Agile Supply Chain: Competing in Volatile Markets", Industrial Marketing Management, Vol. 29, No. 1, pp. 37–44.

Christopher M. (1998), "Logistics and Supply Chain Management: Strategies for Reducing Costs and Improving Services", Pitman, London.

Christopher M., Towill D. (2000), "Supply Chain Migration from lean to agile and customised", Supply Chain Management, Vol 5, No. 4, pp. 206–213.

Cox A., Sanderson J., Watson G. (2000), "Power Regimes: Mapping the DNA of Business and Supply Chain Relationships", Earlsgate Press, Stratford-upon-Avon.

Dekkers R. (2000), "Decision models for outsourcing and core competencies in manufacturing", International Journal of Production Research, Vol. 38, No. 17, pp. 4085–4096.

Dekkers R., Luttervelt C.A. van (2006), "Industrial Networks: Capturing Changeability?" International Journal of Networks and Virtual Organisations, Vol. 3, No. 1, pp 1–24.

Fine C.H. (1998), "Clockspeed: Winning Industry Control in the Age of Temporary Advantage", Perseus Books, Reading, MA.

Fisher M. (1997), "What is the right supply chain for your product?", Harvard Business Review, Vol. 75, No. 2, pp. 105–117.

Fixson S.K. (2005), "Product architecture assessment: a tool to link product, process, and supply chain design decisions", Journal of Operations Management, Vol. 23, No. 3–4, pp. 345–369.

Forza C., Salvador F., Rungtusanatham M. (2004), "Coordinating product design, process design, and supply chain design decisions Part B. Coordinating approaches, tradeoffs, and future research directions", Journal of Operations Management, Vol. 23, pp. 319–324.

Goldman, S.L., Nagel, R.N. and Preiss, K. (1995), "Agile Competition and Virtual Organisations", Van Nostrand Reinhold, New York.

Harrison T.P. (2001), "Global supply chain design", Information Systems Frontiers, Vol. 3, No. 4, pp. 413–416.

Harrison A., Christopher M., van Hoek R. (1999), "Creating the agile supply chain", School of Management Working Paper, Cranfield University, Cranfield.

Hauser J.R., Clausing D. (1988), "The House of Quality", Harvard Business Review, Vol. 66, No. 3, pp. 63-73.

Henderson J.C., Venkatraman N. (1993), "Strategic Alignment: Leveraging Information Technology for Transforming Organizations", IBM Systems Journal, Vol. 32, No. 1, pp. 4-16.

Huang G.Q. (1996), "Design for X: Concurrent Engineering Imperatives", Chapman & Hall, London.

Hult G.T.M., Swan K.S. (2003), "A research agenda for the nexus of product development and supply chain management processes", Journal of Product Innovation Management Vol. 20, No. 6, pp. 427–429.

Ismail H.S., Snowden S.P., Poolton J, Reid I., Arokiam I.C. (2006) "Agile Manufacturing Framework and Practice", International Journal of Agile Systems and Management, Vol. 1, No. 1, pp. 11–28.

Ismail H.S., Sharifi H. (2005), "Supply Chain Design and Design for Supply Chain: A balanced approach to building agile supply chains", Proceedings of ICAM 2005, Helsinki, July.

Ismail H.S., Arokiam I., Reid I., Poolton J., Mooney J. (2007), "How SME's effectively participate in the mass customisation game", IEEE Transactions on Engineering Management, Vol. 54, No. 1, pp. 86–97.

Joglekar N., Rosenthal R. (2003), "Coordination of design supply chains for bundling physical and software products", Journal of Product Innovation Management, Vol. 20, No. 5, pp. 374–390.

Kidd P. (1994), "Agile Manufacturing: Forging New Frontiers", Addison-Wesley, Reading, MA.

Kehoe D.F., Sharifi H., Dani S., Burns N.D., Backhouse C.J. (2007), "Demand Network Alignment; Modelling the DNA of supply chains", International Journal of Production Research, Vol. 45, No. 5, pp. 1141–1160.

Kehoe D.F., Boughton N.J., Sharifi H. (2002), "The role of e-Business in Demand Network Alignment", the 7th International Symposium on Logistics "Integrating Supply Chains and Internal Operations Through e-Business", Melbourne, Australia, 14–17 July.

Lamming R., Johnsen T., Zheng J., Harland C. (1999), "An initial classification of supply networks", International Journal of Operations and Production Management, Vol. 20, No. 6, pp. 675–691.

Lee H.L., Sasser M.M. (1995), "Product universality and design for supply chain management", Production Planning and Control, Vol. 6, No. 3, pp. 270-277.

Lee W.B, Lau H.C.W. (1999), "Factory on demand: the shaping of an agile production network", International Journal of Agile Management Systems, Vol. 1, No. 2, pp. 83–87.

Luftman J.X., Lewis P.R., Oldach S.H. (1993), "Transforming the Enterprise: The Strategic Alignment of Business and Information Technology Strategies." IBM Systems Journal, Vol. 32, No. 1, pp. 198–221.

Lummus R., Vokurka R. (1999), "Defining supply chain management: a historical perspective and practical guidelines", Industrial Management & Data Systems, Vol. 9, No.1, pp. 11–17.

MacMillan I.C., McGarth R.G. (1996), "Discover Your Product's Hidden Potential", Harvard Business Review, Vol. 74,No. 3, pp. 58–73.

Miltenburg J. (1995), "Manufacturing Strategy: How to formulate and Implement a Winning Plan", Productivity Press, Portland, OR.

Nagel R., Dove R. (1993), "21st Century Manufacturing. Enterprise Strategy", Iacocca Institute, Lehigh University, Bethlehem, PA.

Petersen K.J., Handfield R.B., Ragatz G.L. (2005), "Supplier integration into new product development: coordinating product, process and supply chain", Journal of Operations Management, Vol. 23, No. 3/4, pp. 371–88.

Poolton J., Ismail H.S., Reid I.R., Arokiam C. (2006), "Agile Marketing for the Manufacturing Base SME", Marketing Intelligence & Planning, Vol. 24, No.7, pp. 681–693.

Sharifi H., Kehoe D.F., Boughton N.J., Michaelides Z., Burns N.D., Dani. S. (2002) "e-business models in the support of demand networks alignment", Proceedings of the POM 2002 conference, April 5-8, San Francisco.

Sharifi H., Zhang Z. (1999), "A methodology for achieving agility in a manufacturing organisation: an introduction", International Journal of Production Economics, Vol. 62, No. 1-2, pp. 7–22.

Sharifi H, Ismail H.S, Reid I. (2006), "Achieving agility in supply chain through simultaneous "design of" and "design for" supply chain"", Journal of Manufacturing Technology Management, Vol. 1, No. 8, pp. 1078–1098.

Svensson G. (2000), "A conceptual framework for the analysis of vulnerability in supply chains", International Journal of Physical Distribution & Logistics Management, Vol. 30, No. 9, pp. 731–50.

Van der Vorst J., Beulens A. (2002), "Identifying sources of uncertainty to generate supply chain redesign strategies", International Journal of Physical Distribution & Logistics Management, Vol. 32, No. 6, pp. 409–430.

Van Hoek, R., Harrison A., Christopher M. (2001), "Measuring agility capabilities in the supply chain", International Journal of Operations & Production Management, Vol. 21, No. 1/2, pp. 126–147.

Voss C., Tsikriktsis N., Frolich M. (2002), "Case research in Operations Management", International Journal of Operations & Production Management, Vol. 22, No. 2, pp. 195–219.

Yin R. (1994), "Case Study Research", Sage Publication, Beverly Hills, CA.

Zhang Z., Sharifi H. (2007), "Towards Theory Building in Agile Manufacturing Strategy—A Taxonomical Approach", IEEE Transactions on Engineering Management Journal, Vol. 54, No. 2, pp. 351–370.

PART III:

International Issues of Industrial Networks

Control and Coordination for Loosely Connected Networks

The Chapters 6 and 7 offer a perspective on how control and coordination can take place. Both emphasise the design of the supply chain, albeit from different points of departure. The focus of Chapter 6 is on the agile supply chain consisting of more loosely connected entities although not fully accounting for its "chaotic" behaviour; the ASDN open source software enables the coordination in these networks. This work can be viewed as an elaboration of thoughts proposed by Gunasekaran (1998) and as an answer to the call of Sanchez & Nagi (2001, pp. 3594-3595) for tools and information systems for coordination and planning in agile manufacturing systems. This applies to Chapter 7, too, where the significance is found more in the design of agile supply chains; this chapter has related that to the growth and diversification strategy of Ansoff. Both chapters offer a step towards resolving the issue on whether a supply chain should be agile or lean (Towill & Christopher, 2002); Chapter 6 links this to modularity and the Order Entry Point (or Order Decoupling Point), see also Dekkers (2006). So far, the chapters in this book have more or less accounted for loosely connected entities that collaborate, but how about the international issues that arise from these networks?

International Issues of Industrial Networks

The final contributions in Part III touch on the international aspects of industrial networks due to geographical dispersion. Shi (2003) defines three main challenges for international manufacturing networks: the manufacturing value creation process, the internationalisation process, and the inter-firm collaboration formation process. That the internationalisation process leads to more loosely connected entities is underlined by Ernst (1997, p. 101). In that perspective, the contribution by Joachim Kuhn expands on the evolution of the automotive industry that created Dispersed Manufacturing Networks, to reach local markets and to battle competitive pressures by reducing cost (Chapter 8). Stephen Smith *et al.* build on this proposition by investigating to what extent companies have geographically dispersed their manufacturing capacity (Chapter 9). After these two chapters that address mostly the value creation process and the internationalisation process, the contribution by Mahendrawathi Er & Bart MacCarthy (Chapter 10) shows these effects of globalisation on supply chains by the simulation model they have developed. Finally, in Chapter 11, Harsh Karandikar offers a unique model for managing the transition of global engineering networks, moving away from engineering management based on the monolithic company; that chapter addresses the intra-firm collaboration process as a result of dispersion. These contributions align with the need for improved decision-making at managerial level while proving that operational issues will determine the success of managing dispersed manufacturing or for smaller firms the participation in loosely connected networks spanning the globe.

References

Dekkers, R. (2006) Engineering Management and the Order Entry Point, International Journal of Production Research, Vol. 44, No. 18-19, pp. 4011–4025.

Ernst, D. (1997) From Partial to Systemic Globalization: International Production Networks in the Electronics Industry, Berkeley Roundtable on the International Economy, Working Paper 98, Berkeley.

Gunasekaran, A. (1998) Agile manufacturing: enablers and an implementation framework, International Journal of Production Research, Vol. 36, No. 5, pp. 1223–1247.

Sanchez, L.M. and Nagi, R. (2001) A review of agile manufacturing systems, International Journal of Production Research, Vol. 39, No. 16, pp. 3561–3600.

Shi, Y. (2003) Internationalisation and evolution of manufacturing systems: classic process models, new industrial issues, and academic challenges, Integrated Manufacturing Systems, Vol. 14, No. 4, pp. 357–368.

Towill, D. and Christopher, M. (2002) The Supply Chain Strategy Conundrum: To Be Lean Or Agile Or To Be Lean And Agile, International Journal of Logistics: Research and Applications, Vol. 5, No. 3, pp. 299–309.

8 Developing a Worldwide Production Network

Joachim Kuhn
Anglia Ruskin University
Daimler AG

Abstract Creating a worldwide network of manufacturing locations is one of the key issues of globalization. In bundling all these locations together a strategic approach is necessary to generate sustainable network structures. However, a worldwide production network becomes alive only if the strategy can be linked with operational reality. This chapter focuses on this link by forming strategic-operational equations to support the decision-making concerning which manufacturing units are more profitable within a changing and developing network. Based on this quantitative outcome, qualitative aspects complement the investment decision. Due to tremendous cost pressures third party manufacturing (3PM) in the automotive business may improve the company's competitive cost position – under the premise that several Original Equipment Manufacturers' production as well as product technologies/techniques can be merged.

Keywords Automotive manufacturing, Globalization, Production network

8.1 Introduction

When Henry Ford put Model T on the market in the beginning of the 20th century, he had built up a fully integrated plant for mass manufacturing of this car in Dearborn near Detroit: from iron ore processing until final assembly every production step was within the factory site of Ford. Only the transportation of these vehicles to remote locations of the USA proved to be difficult as the shipment of the completely built-up units (CBU) was associated with a high cost burden and sometimes led to quality problems. The solution emerged when the Ford Automotive Company in Dearborn began to supplement CBU car deliveries with deliveries of car kits to the different branches. These kits consisted of parts or sub-assembled modules as determined by the knocked down (KD) degree. The KD kits or CBU cars were transported by train, trucks or ships. Transportation cost was a significant factor in the decision whether to ship the Model T as CBU or KD. The shipment of vehicle kits was a domestic North American transport optimization solution for increasing efficiency, as shown in Figure 8.1. and 8.2.

Besides being transported in the USA, the kits were also supplied to overseas branches or plants, such as Ford Australia, India, Ceylon, Burma, Malaya, South Africa or England (Wilkins and Hill, 1964, p. 44), see Figure 8.3. All Original Equipment Manufacturers (OEMs) in automotive manufacturing have implemented this form of shipping vehicles to other destinations, particularly those abroad.

Figure 8.1. *Shipment of KD kits (Ford Times, 1913, p. 15; The Henry Ford: G35053)*

Currently, the KD-kits are supplied to assembly plants which produce up to 10,000 vehicles per year.

However, many of these production branches/plants were built up based on tariff and non-tariff import barriers (Moran and Riesenberger, 1994, pp. 34–36). By having this visible hand of the state automotive companies could be attracted to make direct foreign investments in KD-supplied factories being profitable only by tariff incentives. With the changing of the political and economic framework to a more liberal one, tariff-based promoted plant locations in many cases became inefficient, with the result that KD-delivered plants started being closed down and CBU-vehicles were imported. For example, in Switzerland 1936 was the time to have the very first KD-plant opened by General Motors (Guenther, 2003). The changed legal framework for customs allowed KD imports and local assembly due to a favourable customs rate compared to CBU cars. This production facility stopped manufacturing cars in 1975 because Switzerland ratified the EFTA (European Free

Figure 8.2. *Shipment of bodies by truck (Ford Times, 1914, p. 7; The Henry Ford: G35055)*

Figure 8.3. *Supply of KD-kits to South Africa (Ford Times, 1913, p. 16; The Henry Ford: G35054)*

Trade Association) Agreement and effected an association agreement with the predecessor institution of the European Union: the favourable customs regulations for importing KD kits were lifted, which had the consequence that CBU imports came to replace the local assembly of vehicles.

From the point of view of nations the question arises whether ex post facto a sustainable change in automotive and related spin-off industry structures has been achieved with the help of trade barriers. The answer is clustered in four main research areas:

1. Implications of trade policy to attract new business with the example of the automotive sector in selected nations.
2. Factors impacting the decision on foreign direct investment in the automotive business.
3. Evolution of production networks on a worldwide scale.
4. Sustainable automotive production structures and their spin-offs in selected nations after the lifting of trade barriers.

This chapter focuses on the evolution of production networks both in strategic/operational perspective and the strategic limits of a borderless networking in the automotive sector.

8.2 Production Structures in the Automotive Sector

8.2.1 Strategic Perspectives of Globalized Production

Setting up a strategy for coordination of all activities related to manufacturing, enables a company to determine its operational structures and processes for production. In the first instance, a strategy, which is either knowledge-, resource- or capability-based (Mueller-Stewens and Lechner, 2003, pp. 356–364) is developed on a business and corporate level. The generic strategic approaches for these two levels are represented by competitive product-related strategies (cost leadership, differentiation or a focus in particular segments (Porter 1980, pp. 34–46)) or technological strategies (technology leadership, [fast] follower, application engineering or me-too [Ansoff, 1990, pp. 43–47; Ansoff and Steward, 1967, pp. 71–83]). The business/corporate strategy provides the basis for derivation of all functional strategies, e.g. for purchasing, logistics, quality or production (Hahn, 1997, pp. 151–153; Hinterhuber, 1996, pp. 44–47; Dekkers, 2003, pp. 386f.). Picking out the production of goods and services it is a potential source for competitive advantage since it fixes a great part of the company's cost burden, the quality level of the product and the speed of product availability to the customer. The latest globalization wave has brought a third strategy pillar into focus: A company has to define how it goes global – either ethno-, poly-, geo- or eclectocentrically (Perlmutter, 1969; Dunning, 1993; Kuhn, 2000) – and which role (black hole, implementer, contributor or strategic leader [Bartlett and Ghoshal, 1986]) is assigned to existing and potential units – thus marking the emergence of a long-term manufacturing network.

In considering the automotive industry and focussing on the manufacturing of a passenger car vehicle, the process is split into four stages: Press shop, body shop, paint shop and final assembly. It starts in the press shop where the coils (one long metal sheet rolled like a household paper towel) are separated into single sheets and then pressed into the different shapes of interior and exterior panels like front fenders or boot lid. In the next stage the body parts needed for a car are either welded or glued together forming the pure body of the car. After washing the body different processes like cataphoretic grounding for corrosion protection, sealing of the defined body sections against incoming water, and final painting are implemented to obtain the painted body. This body is then forwarded to final assembly, where most of the parts like steering, engine, seats, front and rear bumper, wheels, etc., are fitted into the car. With its final release at the last station of the assembly line the car gets its "o.k."-sticker and leaves the factory hall. Subsequently, the vehicle gets routed into sales for distributing it to the final customer – once the car is made to order.

With regard to the KD supply in the automotive industry, certain patterns have evolved, such as, for example, the lot size of 6 or respective multiples thereof for the packing of KD-units. These kits – comprised of app. 1,500 parts (average part amount for Mercedes-Benz passenger cars) – are packed in cardboard boxes or in wooden crates to be sent abroad. Upon their arrival at the destination (e.g. Thailand) the foreign plant unpacks the parts set and builds the cars: at a minimum with the final assembly and painting (semi-knocked down: SKD) and at the maximum with the body shop included (completely knocked down: CKD). The share of knocked down parts depends on tariff (e.g. import duties or excise tax) and/or non-tariff (e.g. local content regulations or quotas) barriers for importing cars/car kits to a particular country or economic region.

Customs regulations are one of the most important levers for control of the incoming CBU car and KD kit quantities from a national point of view. Developed during the 1920s/1930s import custom duties were raised to attract OEMs to invest into local plant development– particularly in Europe and between Europe and the USA. The import duty difference between a CBU car and a CKD-kit was, and continues to be, the factor that triggers the initial impetus for local investments. As at July 2006 for Thailand the tax difference is 80% (CBU import) to 30% (CKD kit import tax) resulting in an investment allowance that is expressed as

$$\text{Investment allowance}_{\text{Customs}} = \text{Tax}_{\text{CKD}} * \text{Import volume}_{\text{CKD}} - \text{Tax}_{\text{CBU}} * \text{Import volume}_{\text{CBU}}$$

with the assumption that an OEM is not supplying the same vehicle as CBU and CKD-kit car to the market.

During the industrialization boom in the 1920s/1930s, when the automobile industry in Europe became a key player, countries around the world grasped the chance to foster local (national) manufacturing locations. A few decades later, cascading to other nations has also become a common scheme in Asia, China, Africa and other developing countries/regions. Thailand, for example, had strong local content regulations. The relevant document – generally enacted by the Ministry of Industry (Automotive Industry Development Committee) – specified the mandatory and voluntary scopes of vehicle parts to be produced locally (Factory Act, 1992). The 1994 version stipulated that parts such as the radiator, chassis wiring harness, wheels, windshield, assembly of the engine, centre floor parts and other parts/subassemblies have to be made in Thailand. The selection of parts was set by the Ministry, whose decision was guided by already existing industries and the strategic outlook for the industrial/service sectors favoured by the Thai Government. Each selected item was assigned a fixed local content percentage, which was to be summed up to at least 54%. This total local content percentage was revised on a continuous basis in accordance with the speed of industrialization. Due to the Asian crisis in the late 1990s and the General Agreement on Tariffs and Trade (GATT) these local content restrictions were lifted, so that, currently, no mandatory local content is in existence for Thailand. With the upcoming Asian Free Trade Agreement (AFTA) the local content requirements will be revitalized. Only if one nation has signed the AFTA it gains 40% local content and then quasi free exports to other AFTA nations (only a minimum import duty of 5% is set) are possible for exchanging goods/services between the relevant nations. Another example is South Africa. During the 1960s and 1970s the degree of local content was measured by weight: 40% of the total weight of a car was mandatory as local content but there was freedom of choice for the OEM to define which parts are affected by the local content provisions. In many cases, the engine or body parts were targeted for localization in support of the heavy industry in the Republic of South Africa. After lifting that local content regulation the South African Government decided to launch a national industrialization programme led by a task force of the motor industry. The Motor Industry Development Programme (MIDP) Phase VI (Ministry of Trade and Industry, 1995), which was enforced in 1995, abolished any local content but instituted a compensating import/export duty: As a kind of economic incentive to produce goods locally, the import duty for CBU cars and CKD kits is balanced against the export rebates for parts, manufacturing components or tools made in South Africa. The South African Government is confident that the motor industry in particular can compete on an international level but still needs some supportive actions like the rebate scheme to initialize a (global) sourcing of goods from South Africa. Success stories include leather interior parts (e.g. seat covering) or metal parts for car bodies – both parts being supplied by local companies to different OEMs worldwide.

As the world has opened up for the free flow of business between regions, the already existing manufacturing facilities and sales units are pushed into global competition. For positioning the different manufacturing units in one global network, a set of alternatives is available in terms of business/corporate strategy and roles assigned to the units. From a strategic point of view, companies can choose a scheme for mastering their own globalization from four basic types (Perlmutter, 1969; Dunning, 1993; Kuhn, 2000), see Figure 8.4:

- *Ethnocentric type.* The company is focussing on the home market and therefore the products are standardized and not modified for other regional markets. Hence, the company mainly exports its products. Only if local content regulations are in force do CKD-delivered factories exist localizing the body and paint shop as well as the final assembly. Depending on the local content regulations, some parts and related productions like assembling these parts are also localized. Hence local units are only in existence if political constraints like tariff or non-tariff barriers are in effect. Once these vanish the local units are given up for the most part.
- *Polycentric type.* Products are tailored to multiple domestic (local) markets and the pursued strategic view is called multidomestic. The product itself only gets minor changes to enable continuation of uniform branding respecting certain key messages. Most of the production volume comes from the core plants but also CKD-set exports – reduced by local content parts – are given a bigger share in companies' production figures. As in the ethnocentric scheme political constraints determine the volume of local investments. However, due to higher

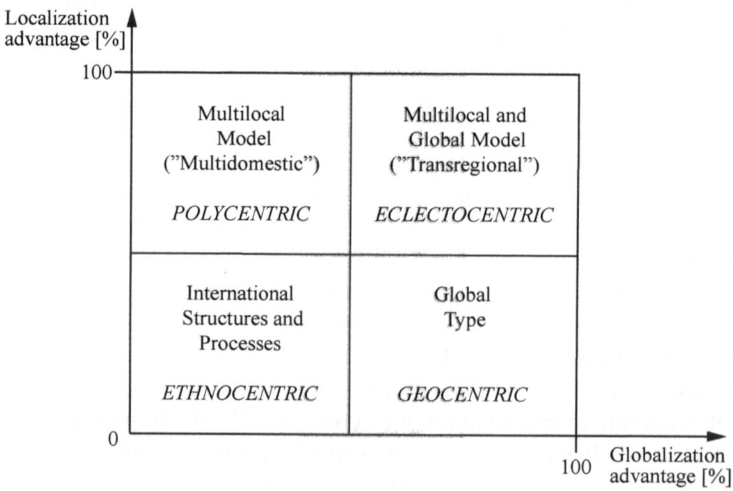

Figure 8.4. *Generic attitudes to globalize*

demand figures the economic benefit supports a local approach even when trade barriers are diminishing.

- *Geocentric model.* The holding company integrates a variety of transplants which operate quite independently from the core plants. Each production site has its own localization strategy, aimed at active local content management: Legal forces as well as economic evaluations determine the local parts production. Hence economic benefits are a major cornerstone in justifying investments in local manufacturing and/or service units that could result in stable and lasting structures.
- *Eclectocentric type.* The company is choosing between the poly- and geocentric model influenced by the regional market conditions. Plant structures are CKD-ones or transplant orientated with both plant types depending on two to three core factories. The focus of each plant is to produce particular products and parts or modules bound for a regional sale or global transfer between the different regions creating a transregional strategy. Since investments are both political- and/or economy-driven the same consequences apply as for the poly- and/or geocentric view: The sustainability of investments depends on the country-specific circumstances.

After selecting the appropriate strategy, management has to determine the individual assignment of roles within the network (Ghoshal and Bartlett, 1998, pp. 121–130). Focusing on production units it involves local resources, knowledge and capabilities to elaborate one of the following four principal roles expressing the operational strategy (see Figure 8.5):

- *Implementer.* The task of the regional company is to implement the core plants' methodologies and facilities. The primary target of the subsidiary is to provide goods only to the local market. The production is commonly determined by CKD-kits and the particular local content productions. The access to core-company-related critical knowledge, the latest technologies, as well as the financial scope, are limited due to the above mentioned target. In many cases these local factories are considered as "extended work benches" of the core plants or transplants.

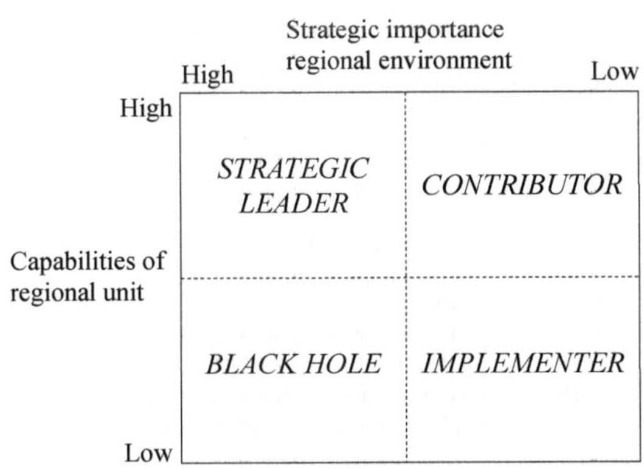

Figure 8.5. *Generic roles of company units*

- *Contributor.* The regional companies are not only seen as facilities for production and sale of goods but also contribute to specific R&D activities in the whole corporation. Hence the local knowledge generation and dissemination engenders the regional capability in particular fields. It also has an influence on the group's and company's strategy. The commonly used factory form is the transplant and in rare cases a CKD-one if the market volume does not allow the investment into a transplant.
- *Strategic Leader.* The regional transplant or core factory has already developed the competencies and required capabilities in a particular technological or product field. All plant activities are covering the fields from censoring weak/strong signals from outside the company (Ansoff 1990, pp. 383–399) until having transformed this information into new technologies and products. The local company is fully responsible for the concerned strategies and its strategy represents the leading edge in corporate strategy formulation.
- *Black Hole.* The local company does not have a significant role because it lacks the required capabilities to have a major impact in regional competition or because a strategic mismatch appears between the headquarters and the local company. In this temporary domain of unawareness ("black hole") the subsidiary/associated company is in the unlucky situation either of having to be bailed out by immense knowledge transfers and other enablers or of being closed down. Therefore, it needs a quick decision by the OEM to redirect the factory into the "normal" roles like an implementer, contributor or strategic leader if the local company is to survive. In many cases the concerned plants are CKD ones in the time periods after the calculated plant life cycle.

By allocating one predominant role to a particular local unit it expresses the status of this location. After reconciliation with the globalization attitude as well as the

Unit's Role / Strategic attitude	Implementer	Contributor	Strategic Leader	Black Hole
ETHNOCENTRIC	1.00	0.50	0.00	0.25
POLYCENTRIC	1.00	0.75	0.00	0.25
ECLECTOCENTRIC	0.25	0.50	0.50	0.25
GEOCENTRIC	0.25	0.75	1.00	0.00

with values expressing the fitness τ

0.00 = total strategic-operational misfit
0.25 = misfit between company's globalization and role of units
0.50 = partly fit
0.75 = favourable fit and economical feasible
1.00 = total fit and economical success

Figure 8.6. *Relation matrix between strategic attitudes and the generic roles of company units*

cooperation between roles of all individual units around the globe the specific role gets approved by the company's strategy board. There are certain correlations obvious to optimise the fit between the strategic global approach and its operationalization. The relationship values τ in Figure 8.6. are set by the author as an example – they have to be determined by each company individually. The value τ expresses the fitness between the two factors. With the best being 1 and the worst being 0 this parameter is used to strengthen or weaken the overall globalization approach. τ is embedded in the calculation for ε as computed by

$$\varepsilon = (\alpha + \beta) \times \tau$$

with ε being the accelerator factor for strengthening or weakening the investment decision in total. α and β are derived from Figures 8.4. and 8.5 and shown in Figure 8.7. The values given are set by the author with the assumption that the enterprise is striving for free trade as set by the World Trade Organisation (WTO, 1999; Croome, 1999). The geocentric attitude in combination with disseminated manufacturing units becoming each a worldwide hub for a particular product range is favoured to be the most suitable approach for globalization.

Generally α and β express the fit (misfit) between company-internal strategic options to globalize and the political expectations as determined by the General Agreement on Tariffs and Trade. The different values of α and β:

0.5 = misfit and political not favourable
1.0 = misfit and political feasible
1.5 = fit and economical/political feasible
2.0 = fit and economical/political favourable

express the degree of fitness. τ outlines the value of the management's choice for a specific combination of attitudes to roles. By implementing these two factors both the internal view of the company and the external political expectations are taken into account. ε represents the strategic level to do the right things and it accelerates the investment amount with

$$\text{Investment allowance}_{\text{Static}} - \varepsilon * \left(\text{Investment allowance}_{\text{Customs}} + \text{Investment allowance}_{\text{Cost / Earnings}}\right)$$

Strategic Attitude

$\alpha = 1.0$ *POLYCENTRIC*	$\alpha = 1.0$ *ECLECTOCENTRIC*
$\alpha = 0.5$ *ETHNOCENTRIC*	$\alpha = 2.0$ *GEOCENTRIC*

Unit's Role

$\beta = 1.0$ *STRATEGIC LEADER*	$\beta = 1.5$ *CONTRIBUTOR*
$\beta = 0.5$ *BLACK HOLE*	$\beta = 2.0$ *IMPLEMENTER*

Notes:
α = Strategy factor
β = Role factor
with α and $\beta = \{0.5, 1.0, 1.5, 2.0\}$

Figure 8.7. *Strategy factor α and role factor β*

where the total static investment allowance is calculated by multiplying the strategic accelerator ε with the investment allowances due to the company's operations: the allowance due to customs is separated because of the political constraints influencing this lump sum. The allowance arising from cost/earnings differences mirrors the company's internal capability to achieve an efficient operation and satisfy the customer with its solution in terms of the product, services and information given (Goldman *et al.*, 1994, pp. 75–83).

8.2.2 Operational Design of Globally Disseminated Manufacturing Units

Processes related to cost/earnings are clustered into several classifications as determined by the value chain (Porter, 1985, pp. 36–53):

I. Primary processes
   - I.1. Inbound/outbound logistics
   - I.2. Production
   - I.3. Marketing and Sales
   - I.4. Services

II. Support processes
   - II.1. Procurement
   - II.2. Technology
   - II.3. Management
   - II.4. Human Resource Management.

In calculating the benefits of a foreign location three factors are taken into consideration:

1 Cost/Earnings.
2 Quality.
3 Time.

Each factor is evaluated via cost equations. Particularly, Production/Logistics (PL) for primary processes and Management cost (M) for support processes significantly influence the cost calculation pro or contra a foreign direct investment as empirical analysis indicate. But beyond process efficiency quality also matters: Comparing the foreign output either on a manufacturing or service level the quality philosophy of *"making it right the first time"* (Garvin, 1988, p. 44) plays an important role when comparing different locations on a yield basis (Shafer and Meredith, 1998, p. 322; Kaplan, 1990, p. 34; Fox, 1994, pp. 303–313). Therefore the cost of non-conformance – controlled as the First Yield Rate (FYR) – are measured. A further evaluation of the whole process network is given by lead time. Besides other time-based instruments (Kuhn, 1995) such as the Market Timing (MT; first vs. follower market entry) or the learning curve coupled with the experience curve the lead time is predominantly controlled and measured by tied-up capital. Thereof quality and time are supporting the traditional cost/earnings comparison.

8.2.2.1 Cost/Earnings Figures

The calculation for the *Investment allowance*$_{Cost/Earnings}$ inherits the factors quality and time and is given by

$$\text{Investment allowance}_{\text{Cost / Earnings}} = \text{Cost allowance}_{\text{Localplant}} + \text{Cost allowance}_{\text{FYR}} + \text{Earnings allowance}_{\text{MT}}$$

with

$$\text{Cost allowance}_{\text{Localplant}} = \lambda * \left[\left(\text{Cost}^{\text{PL}}_{\text{Localplant}} - \text{Cost}^{\text{PL}}_{\text{Alternativeplant}}\right) + \left(\text{Cost}^{\text{M}}_{\text{Localplant}} - \text{Cost}^{\text{M}}_{\text{Alternativeplant}}\right)\right]$$

with λ = expected labour intensity. λ serves as a potential de-/accelerator for the cost allowance with λ equalling

0.5: labour proportion is lower than automation;
1.0: labour and capital are used in approximately the same proportion;
1.5: labour proportion is higher than automation.

The labour intensity reflects the proportion of labour (e.g. the sewing of seat covers and the final assembly of the completed seat into the car) to capital-intensive processes like the body shop with lots of fully automated welding robots. The expected intensity covers future developments and whether the concerned production processes will be automated or given to a manual operation as shown in Figure 8.8.

The formula itself is based on a fixed production volume as a reference for calculating the relevant cost burden for either the local plant or the additional burden for the core factory. The volume produced and sold to the market is a major driving force in calculating the overall cost. The result is split into three areas:

- $\text{Cost allowance}_{\text{Local plant}} < 0$: Advantage for local plant;
- $\text{Cost allowance}_{\text{Local plant}} = 0$: No influence on location decision;
- $\text{Cost allowance}_{\text{Local plant}} > 0$: Disadvantage for local manufacturing unit.

8.2.2.2 Quality Related figures

The *Cost allowance*$_{First\ Yield\ Rate}$ is expressed as

$$Cost\ allowance_{FYR_{Local\ plant}} = \sum_{i=1}^{n}\left(\frac{non-conforming\ cost\ i}{unit}\right) * \left(1 - \frac{FYR}{100}\right) * production\ volume$$

with

Figure 8.8. *Manual vs. automated operation in vehicle manufacturing (example rear screen assembly)*

$$FYR = \frac{Number\ of\ o.k. - units}{Total\ production\ volume} * 100 \quad [\%]$$

and cost classifications *i* as for example labour and material cost for rework, or scrap cost. The formula is calculated for a local and an alternative plant and both values compared as shown in

$$Cost\ allowance_{FYR} = Cost_{FYR\,Local\ plant} - Cost_{FYR\,Alternative\ plant}$$

with having three cases:

- $\text{Cost allowance}_{\text{FYR}} < 0$: Advantage for local plant;
- $\text{Cost allowance}_{\text{FYR}} = 0$: No influence on location decision;
- $\text{Cost allowance}_{\text{FYR}} > 0$: Disadvantage for local manufacturing unit.

8.2.2.3 Time Related Figures

The $Earnings\ Allowance_{Market\ timing}$ is represented in the form

$$Earnings\ allowance_{Market\ timing} = Rapid\ response\ surplus_{CKD\,Local\ plant} - Sales\ volume_{CBU_{Alternative\ plant}}$$

with

$$Rapid\ response\ surplus_{CKD\,Local\ plant} = \eta * \left[\left(\frac{Earnings}{car}\right) * Sales\ volume\ CKD\ cars\right]$$

where η = insiderization accelerator (Ohmae, 1989) with three values given by

- $\eta = 0.5$: local manufacturing bears a low status;
- $\eta = 1.0$: imported CBU cars and local manufactured ones are treated equal;
- $\eta = 1.5$: local manufacturing is honoured by customers (high status).

under the assumption that the import and sales volume match, i.e. no stock on hand is visible. Therefore the volume figures above and the figures from the $Investment\ allowance_{Customs}$ are based on the same numbers.

8.2.2.4 Investment Benefit

Future cost/earnings are calculated by using a modified net present value method (Horngren *et al.*, 1997, pp. 783–788):

$$Investment\ benefit = \sum_{j}^{m} \frac{Investment\ allowance_{Static_{Period\ j}}}{(1+r)^2}$$

with

r = minimum acceptable rate of return ("hurdle rate") and

j = annual period with $j = \{1, 2, \ldots, 10\}$ since strategic planning is 10 years.

The investment benefit can obtain three stages

- Investment benefit > 0 : financial profit.
- Investment benefit $= 0$: no monetary gain.
- Investment benefit < 0 : financial loss.

The different equations represent the investment decision for constructing one single factory in a particular country. Deciding about several alternative locations or a whole network of factories the equations are expanded by the number of potential plant investments. If, for example, three plants are to be realized the formula is modified by adding *k* with $k = \{1, 2, 3\}$ indicating the particular plant as e.g.

$$Cost\ allowance_{Local\ plant\ k} = \lambda_k * \left[\left(Cost^{PL_k}_{Local\ plant_k} - Cost^{PL}_{Alternative\ plant} \right) + \left(Cost^{M_k}_{Local\ plant_k} - Cost^{M}_{Alternative\ plant} \right) \right]$$

All cost/earnings equations help to determine a temporary solution for allocating production volume to manufacturing units. But this network of worldwide dispersed factories needs a production strategy to limit a constantly changing allocation of production figures being in contrast to the *"production smoothing [which] is the cornerstone of the Toyota production system"* (Monden, 1994, p. 8).

8.2.3 Strategic Borderlines for Dynamic Networking

Future networking needs to orientate the different worldwide dispersed locations with one principal idea. As customer-orientation is the key issue in many industrial or service sectors this idea is placed as a nucleus for all other company activities. By emphasizing the customer one of the major challenges has been agility (Goldman *et al.*, 1994). Taking a closer look how to be agile the different production units need to be repositioned as an agile entity (*ibid*, pp. 71-120).

By enriching the customer a company generates not only products but solutions to customers. These context-related solutions consist of the product itself, services regarding the product sale as well as the ongoing customer use, and information about both. All three of them are tailored to an individual customer. This approach also includes the design and engineering of the product: it is a co-design between company and customer. Such a co-evolution of a product is available in the automotive industry to a certain degree. Particularly the number of options and individual design like specific leather colouring or body panel paint with 3D-effect express the customization of a mass product. Tuning companies bring an additional momentum by offering a wide range of parts and accessories. A unique tailoring of a car in the interior and/or exterior design plus an adaptation in the car body itself (for example a long version) or engine changes are met by specific companies nowadays (e.g. for Mercedes-Benz passenger cars: AMG or Brabus) but it already had its roots in the 1920s/1930s. In these times the bodywork, painting and interior were done by specialized manufacturers like Spohn, Auer or Binz. The customer was in a position to determine the whole car in all its details guided by standard modules (e.g. for the body to be built like a Limousine, Phaeton, Landaulet, Roadster or Cabriolet [Wolff Metternich, 1990]). The other side of such enrichment is indicated by an extremely high price. For example, Maybach cars with mainly Spohn-bodies were only made from 1922 to 1941 with a total production of approximately 2390 cars (still existing today: 149 cars [Wolff Metternich, 1996, p. 17]).

In terms of the Order Decoupling Point the Engineer-To-Order (ETO) (Olhager, 2003, pp. 320–324) approach represents this kind of vehicle production. But in the automotive industry – passenger cars – the general order entry into production is

Assemble-To-Order (ATO). Especially in Europe the adding of standardized options – like a sunroof, particular alloy rims, certain interior wood designs, seat ventilation and heating, various leather styles or a wide range of exterior and interior colours – plays a major role in positioning a car in the market. Whereas the Japanese or American OEMs show a car quite fully equipped with only some options like sunroof or few outside or interior colours. Such a standard car can be produced by Make-To-Stock (MTS). There, the customer expects his future car to already be in the showroom and that he will be able to take it away immediately. Instead in Europe he is used to ordering his selected car configuration and waiting for his individualized car. In the Sindelfingen plant, production is about to 300,000 cars per year and only 2 to 3 cars out of this are identical in each option. The design of the factory itself mirrors the complexity of the products – either in manufacturing or the relevant administrative processes. As a global player an OEM has to determine a factory on demand (Lee and Lau, 1999) strategy covering both products: the standard one and the individualized one or alter the customer behaviour. Whether MTS or ATO is applied the actual frontier in vehicle production is reached when engineering is affected – the enrichment of the customer is facing its boundary.

The second dimension of an agile enterprise is cooperation with other companies either internally or externally to bring solutions to the market as rapidly and cost-efficiently as possible. The enterprise uses the resources regardless of their location and ownership – the company takes control of them by means of virtual companies, alliances, cross-functional teams, empowerment and the reengineering approach. Many OEMs working in the automotive industry have established a network on their own during the last decades. Exchanges of manufactured modules between the plants and the assignment of one or two products to a single production location are the main characteristics of today's networking. Only a few cases exist with independent external partners to produce a specific section of the value chain – with the overwhelming majority of partnerships in the final assembly area. Press or body shop as well as painting are commonly not suited for an OEM-independent manufacturing as the current technologies and production techniques do not allow synergy effects. Thus, for example, the BMW X5 series, Jeep Grand Cherokee, Mercedes-Benz M-Class, Porsche Cayenne, Volkswagen Touareg or Volvo XC90 are all sports utility vehicles (SUV) but in total these cars cannot be built in one single factory with all four stages of automotive manufacturing. Nevertheless, strategic alliances like those between Volkswagen Group and Porsche have pursued a shared manufacturing approach (joint off-road vehicle production Bratislava: VW Touareg and Porsche Cayenne) to a certain degree with the final mounting of brand-symbolic parts in OEM-specific plants: Volkswagen still in Bratislava and Porsche in Leipzig.

Due to further cost pressures from Asian production locations – in particular China – all OEMs will face a next round to reduce the overall cost burden. An automotive *star alliance* might be a strategic target. Involving a lot of captive capital, especially production is in the focus of substantial cost reductions across all OEMs. Hence various OEM manufacturing units are merged together guided by a 3PM. The 3PM has to ensure a proper implementation of production technologies and quality management tools across the OEM requirements to accommodate the manufacture. A product-related fit is required as indicated by clustering all vehicles into certain categories: Three criteria are available: technology (Sommerlatte and Deschamps, 1985, pp. 49–52; Servatius, 1985, p. 116), customers (Kotler and Keller, 2006, p. 156f.), and products (= vehicle category) to cluster these as shown in Figure 8.9 (similar to Abell [1980, p. 30f]). Selecting one cluster the 3PM forms neutral zones for distinct production operations and in final assembly the finishing line gets separated to give the particular vehicle its brand touch. Only by having such 3PMs in place can the overall production cost burden in high cost countries be tackled and OEMs become competitive against low cost countries.

The manner of organizing to master change and uncertainty is illustrated by flexible, innovation-orientated entities. These loosely coupled individuals, groups, or entire companies are driven by common goals like minimizing the time from concept to cash (Goldman, 1994, p. 101). With volatile and less predictable customer preferences the flexibility of producing solutions is given more attention than in the past, thus pointing to the idea of a factory on demand (Lee and Lau, 1999). Coping with these demand fluctuations an independent 3PM has to build the flexibility into the machinery, people and management. But particularly the tools and jigs along the production process (press, body and paint shop as well as final assembly) in automobile industry are captive capital. The equipment can only be used for one brand (e.g. Mercedes-Benz), in many cases for only one model line (e.g. S-Class), and sometimes for only one type (e.g. S600 long version). The synergy effects are limited within one brand and even less when producing as 3PM. The situation gets more difficult for a 3PM since production technologies are different for a small serial production compared to a mass customized manufacturing (see Figure 8.9). Hence, the quick changeover from one manufacturing location to another one or the temporary adding of a production plant in the case of peak market demands is not a

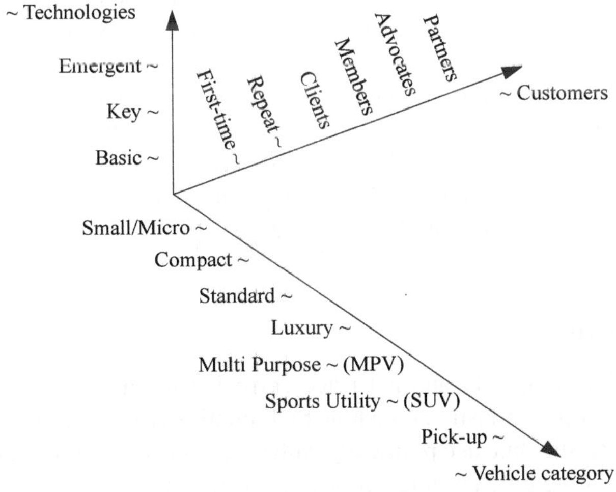

Figure 8.9. *Cluster determinants for allocating vehicle manufacturability groups*

quick fix – as requested by agility. The flexible cooperation of an agile company with other companies is limited in automotive manufacturing.

Leveraging the impact of people and information is the differentiation factor of an agile company to a normal business. The skills, knowledge and information are enablers to create solutions to each customer's individual needs. The skilled workforce is the most valuable "asset" and competitive advantage for an agile company. In the automotive sector the skill level needs to be focused on two sides: one side mirrors the interface between the company and the customer. The better a skilled and knowledgeable sales force promotes solutions, the more convinced the customer becomes and solutions can be processed without any delay within the company. The initial contract does not need any further explanations or inquiries – it is "made right the first time" (Garvin, 1988, p. 44). Once the order is produced the second side comes up: All model options and incremental engineering changes can be built with craftsmanship and not only according to standard work instructions only with a skilled labour force. But the role to mass manufacture cars like craftsmen requires a mind-set to be flexible and know by heart the different options and various car models of some OEM. Only with that brainpower the flexibility and responsiveness both defined as changeability (Dekkers and Luttervelt, 2006, p. 9) is given within minimum time. In particular this mindset and product complexity are key issues if manufacturing is evaluated as a commodity (Bennett and Dekkers, 2005) to be moved or copied quite quickly between worldwide dispersed locations.

Such a craftsmen working style is adapted variously around the globe. Since management used to be "doers" they told their subordinates what to do and how to (Goldman *et al.*, 1994, p. 186). Agility transforms the role into coaching and leading employees to achieve best performance. The distance in power (Hofstede, 1984, pp. 65-109, 257–261; Hofstede, 1994, pp. 139–158) is high in the first case and lower for the latter one, with the manager being more in a role as primus inter pares. By more and more reducing this distance various nations have a significantly altered attitude to incorporate agility. Further research needs to take into account that power distance and other factors like uncertainty and avoidance prove the universality of the agile concept.

Agility is limited in the passenger car industry. In particular, engineering adaptations across OEMs are generally unfeasible because of the missing economies of scale and non-acceptable price levels. Dispersed manufacturing will therefore be the key issue with long-term 3PM-OEM partnerships still being justified by strategic-operational equations as shown above.

8.3 Conclusion

The tremendous overcapacity and net price corrosion as shown, e.g. in the USA, have led the automotive industry as a whole to dramatically redesign itself. In former days not only feasibility but also politically embedded decisions about opening a new factory location formed the argumentation; due to favourable economic conditions set by the government a manufacturing unit had been settled in one country. After the end to these subsidies many of the companies changed location to move on to another favourable location within another nation.

The main contribution of this chapter is directed at rationalizing foreign direct investments in automotive manufacturing units; the plants arising have to be strategically embedded in a corporate as well as functional sense along with an operational justification based on cost/earnings. But it requires strategic borderlines as given by the agile strategy in the automotive sector to ensure a smoothened production. Future research involves the technological standardization of vehicle production across several OEMs. Another research area is opened up by questioning how much customization of a passenger car is really needed and requested by the customer. Since product- and process-related research fields have to be tackled by emerging 3PM or existing OEM the final question comes up: who is going to take the steering wheel?

References

Abell, D. (1980) Defining the Business, Englewood Cliffs: Prentice Hall International.

Ansoff, I. and MCDonnell, E. (1990) Implanting Strategic Management, 2nd edition, New York: Prentice Hall.

Ansoff, I. And Steward, J. (1967) Strategies for a Technology-Based Business, Harvard Business Review, Vol. 45, No. 6, pp. 71–83.

Bartlett, Ch. and Ghoshal S. (1986) Tap Your Subsidiaries for Global Reach, Harvard Business Review, Vol. 64, No. 6, pp. 87–94.

Bennett, D. and Dekkers, R. (2005) Industrial Networks of the Future - A Critical Commentary on Research and Practice, in: Proceedings of the 12th International EurOMA Conference, Budapest.

Croome, J. (1999) Reshaping the World Trading System, 2nd edition, The Hague: Kluwer Law.

Dekkers, R. (2003) Strategic capacity management: meeting technological demands and performing criteria, Journal of Materials Processing Technology, Vol. 139, No. 1-3, pp. 385–393.

Dekkers, R. and van Luttervelt, C. (2006) Industrial networks: capturing changeability?, International Journal Networking and Virtual Organisations, Vol. 3, No. 1, pp. 1-24.

Dunning, E. (1993) Multinational Enterprises and the Global Economy, Addison-Wesley: Wokingham (UK).

Factory Act (1992) Factory Act B.E. 2535 including the Notification of the Motor Industry Development Committee (1994), dated 18th January 1994 thereof including the appendix of compulsory and mandatory vehicle parts/components, Bangkok.

Ford Times (1913) Crating Fords for Export, Ford Times, Vol. 7, No. 1, pp. 15–16.

Ford Times (1914) A Carload of Cars, (Canadian) Ford Times, Vol. 2, No. 1.

Fox, M. (1994) Quality Assurance Management, Reprint of 1994, London: Chapman & Hall.

Garvin, D. (1988) Managing Quality, New York: The Free Press.

Ghoshal, S. and Bartlett, Ch. (1998) Managing Across Borders, 2nd edition, London: Random House.

Goldman, S., Nagel, R. and Preiss, K. (1994) Agile Competitors and Virtual Organizations, New York: van Nostrand Reinhold.

Guenther, D. (2003) Die Schweiz kann mehr als nur Geld und Schokolade [Switzerland is able to do more than money and chocolate], Frankfurter Allgemeine Zeitung, No. 51 of 1st March, p. 49.

Hahn, D. (1997) US-amerikanische Konzepte strategischer Unternehmungsführung [US-American concepts for strategic company management], in: Hahn, D. and Taylor, B. (1997) Strategische Unternehmungsplanung – Strategische Unternehmungsführung [Strategic Company Planning – Strategic Company Management], 7th edition, Heidelberg: Physica Publishing, pp. 144–164.

Hinterhuber, H. (1996) Strategische Unternehmungsführung [Strategic Company Management], 6th edition, Berlin: Walter de Gruyter.

Hofstede, G. (1984) Culture's Consequences, abridged edition, London: SAGE.

Hofstede, G. (1994) Culture and Organizations, London: Harper Collins Business.

Horngren, Ch., Foster, G. and Datar, S. (1997) Cost Accounting, 9th edition, London: Prentice Hall.

Kaplan, R. (1990) Measures for Manufacturing Excellence, Boston: Harvard Business School Press.

Kotler, P. and Keller, K. (2006) Marketing Management, 12th edition, Upper Saddle River: Pearson/Prentice Hall.

Kuhn, J. (1995) Das zeitgesteuerte Unternehmen [The Time-Driven Company], Frankfurt/New York: Campus Publishing.

Kuhn, J. (2000) The Role of Continuous Improvement within Globalization, International Journal of Technology Management, Vol. 20, No. 3/4. pp. 442–458.

Lee, W. and Lau, H. (1999) Factory on demand: the shaping of an agile production network, International Journal of Agile Management Systems, Vol. 1, No. 1/2, pp. 83–87.

Ministry of Trade and Industry (1995) Proposal for a revised Phase VI Motor Development Programme, Cape Town (South Africa).

Monden, Y. (1994) Toyota Production System, 2nd edition, Norcross: Industrial Engineering and Management Press.

Moran, R. and Riesenberger, J. (1994) The Global Challenge, London: McGraw Hill.

Mueller-Stewens G. and Lechner C. (2003) Strategisches Management [Strategic Management], 2nd edition, Stuttgart: Schaeffer-Poeschel Publishing.

Ohmae, K. (1989) Managing in a Borderless World, Harvard Business Review, Vol. 67, No. 3, pp. 152–161.

Olhager, J. (2003) Strategic positioning of the order penetration point, International Journal of Production Economics, Vol. 85, No. 3, pp. 319–329.

Perlmutter, H. (1969) The Tortuous Evolution of the Multinational Corporation, Columbia Journal of World Business, Vol. 4, January/February, pp. 9–18.

Porter, M. (1980) Competitive Strategy, New York: The Free Press.

Porter, M. (1985) Competitive Advantage, New York: The Free Press.

Servatius, H.-G. (1985) Methodik des strategischen Technologie-Managements [Methodology of Strategic Technology-Management], 2nd edition, Berlin: Erich Schmidt.

Shafer, S. and Meredith, J. (1998) Operations Management, New York: Wiley.

Sommerlatte, T. and Deschamps J.-Ph. (1985) Der strategische Einsatz von Technologie – Konzepte und Methoden zur Einbeziehung in der Strategieentwicklung des Unternehmens [The strategic implementation of technology – Concepts and methods to integrate it into the strategy development of the company], in: Little, A. (1985): Management im Zeitalter der strategischen Führung [Management in the Era of Strategic Management], Wiesbaden: Gabler, pp. 37–76.

Wilkins, M. and Hill, F. (1964) American Business Abroad, Ford on Six Continents, Detroit: Wayne State University Press.

Wolff Metternich, M. (1990) Distanz zur Masse [Distance to the Mass], Lorch: Herman E. Sieger.

Wolff Metternich, M. (1996) Maybach Register [Maybach Register], 3rd edition, Lorch: Herman E. Sieger.

WTO (1999) The Legal Texts – The Results of the Uruguay Round of Multilateral Trade Negotiations, Cambridge: Cambridge University Press.

9 Planning in Companies with Dispersed Capacity

Stephen A. Smith
Zen Internet

David J. Petty
Manchester Metropolitan University

David G. Trustrum
Strix Innovative Technologies

Ashraf W. Labib
Portsmouth Business School

Abstract This chapter's objective is to discuss the effect of dispersed capacity on planning in manufacturing organisations. The chapter is based on case study research conducted by the authors in a company with geographically dispersed capacity. This is augmented by a questionnaire survey intended to show that the case study company is not unique. The information collected from the two previously mentioned sources is compared with literature. The questionnaire and literature indicated that there are now substantial numbers of companies with dispersed capacity. The case study research indicates that this leads to greater planning complexity because of increased lead-times due to transhipment. In contrast with single-site companies, this complexity cannot be addressed by Just-in-Time in cases where capacity is dispersed. In-depth research has been conducted in only one company, though the questionnaire survey provides supporting data. The chapter implies that companies employing dispersed capacity need to focus on improving their planning systems to cope with the increased complexity. While the literature on Supply Chain Management (SCM) is extensive, there has been relatively little case-based research on the implications of globalisation for planning. In particular, while it is widely believed that many companies have dispersed their capacity, the evidence for this is largely anecdotal. The survey presented in the chapter provides quantitative data to support this belief.

Keywords: Case study, Logistics, Supply Chain Management, Survey

9.1 Introduction

Supply Chain Management (SCM) has attracted an enormous amount of interest from both academics and practitioners. Examples of key works on Supply Chain Management include: Lambert and Cooper (2000), Harland (1996), and Mason-Jones *et al.* (2000). Distributed Manufacturing has addressed the issue of dispersed manufacturing planning, and scheduling. Examples of such work include: Azevedo and Sousa (2000), Duffie and Prabhu (1996), Maturana, and Norrie (1996), and Shen

(2000). Over the last sixty years, numerous SCM philosophies have been developed for manufacturing organisations. Over this period, however, manufacturing has actually been in decline in Western Europe. In the UK, there is disagreement over the level of the decline of the contribution of manufacturing to the Gross Domestic Product (GDP) because the price of manufactured goods has fallen faster than in other sectors. Certainly, there is a strong perception that manufacturing is in decline in the UK to such an extent that the Department of Trade and Industry (DTI) has proposed measures to improve the image of the sector (DTI, 2004). Across the world, however, manufacturing has grown significantly. Since 1990, Manufacturing exports across the world have grown by 103% (The Globalist, 2005). The majority of this growth has been in the developing nations whose exports have grown by 224% over the same period as opposed to 69% for developed nations. Many western companies are choosing to invest in the developing nations. There has been a total of 9008 foreign direct investment projects in the Asia-Pacific region since 2002 and 5754 in Eastern Europe (Locomonitor, 2005).

There is a large volume of literature on networks and the extended enterprise. Examples of such work include: Frederix (2001), O'Neil and Sackett (1994), and Karlsson (2003). There is evidence for a movement of manufacturing capacity into countries such as India and China. In some cases, this movement may be the wholesale outsourcing of manufacturing to an area with low labour costs. An alternative approach is advocated by Preston (2004) where an established company re-locates some of its capacity offshore, but retains a measure of control and ownership. This is a strategy adopted by Strix Ltd, a leading manufacturer of controls and cordless interfaces for kettles, jugs and a wide range of water boiling appliances. Strix controls are the enabling technology that kettle manufacturers use to give their products extra features. They are designed to work in conjunction with the kettle's heating element, switching the kettle off when the water inside boils. The controls also protect kettles and its user in the event of the kettle boiling dry. Strix Ltd and the University of Manchester have been engaged in collaborative research over the

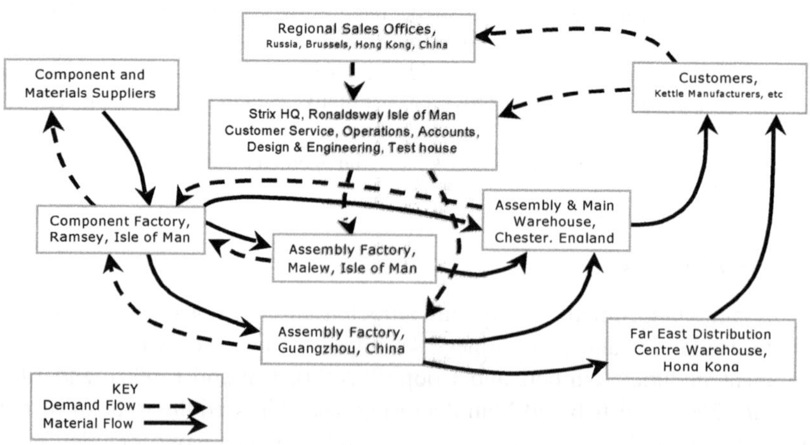

Figure 9.1. *Strix Ltd sites*

last four years in the area of supply chain performance. This chapter is based largely on this research.

The aim of the chapter is to discuss the supply chain problems faced by organisations with capacity that is widely geographically distributed. This discussion will be based on the case of Strix, a review of the key literature and the results of a questionnaire survey.

Therefore, the research methodology is based on Action Research. There are two main limitations associated with action research and these are: (i) reliability and validity of the action research arising from the involvement of the researcher and action research members (see Eden and Huxman, 1996), and (ii) generalisation of the action research resulting from the transfer from the selected case to other cases (see Argyris and Schon, 1989). To overcome these limitations, the authors have triangulated this research through surveys and comparison with related sources in the literature.

9.2 Profile of Strix Ltd

Strix Ltd is based on the Isle of Man (IOM), but now also has manufacturing capacity located in mainland UK and China. Figure 9.1 details the various Strix sites and how demand and materials flow between them.

While Strix has realised significant savings in labour costs, the increased lead times have led to several logistical problems as follows:

- Internal lead times are now longer than the firm demand horizon for the majority of products, forcing production plans to rely on forecasts. As can be seen from

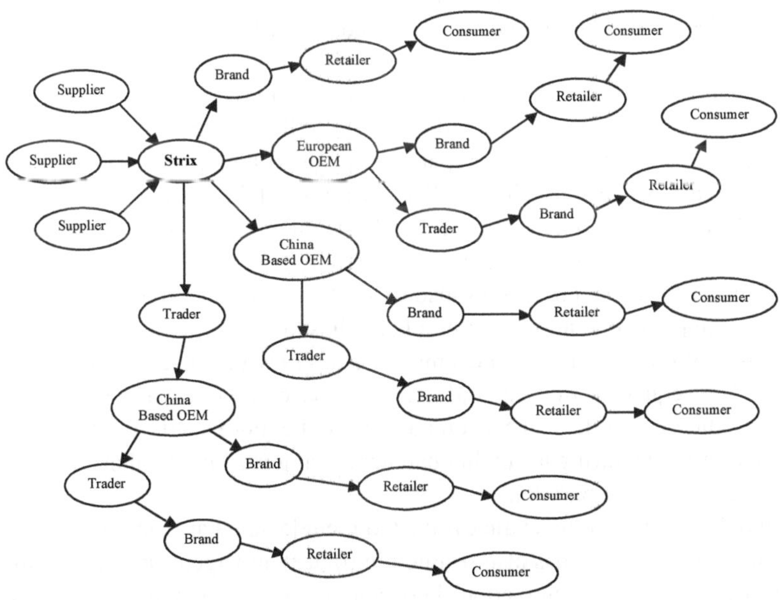

Figure 9.2. *Simplified version of the domestic appliance global supply chain*

Figure 9.1, transhipment of items between the UK and China is required. Typically, the shipping time between the UK and China is six weeks. Air freight is much faster, but this is normally only used in the cases of emergency because of the very high associated transport cost.

- The introduction of wider global markets has resulted in much less predictable patterns in incoming demand. Strix is part of a highly complex supply chain, a simplified representation of which is presented in Figure 9.2.
- The differences in manufacturing styles in different countries make it difficult to visualise the requirements of all concerned and the actions they are likely to take with respect to logistics and supply chain decisions.

There is much anecdotal evidence to suggest that the difficulties facing Strix in terms of market visibility, forecasting and planning are a global issue, but there is no quantitative data to support this view.

The most significant change in the domestic kettle supply chain structure has been the growth of global markets outside Western Europe, in particular the Far East. In 1997, Strix opened its first manufacturing facility in China in response to the growing numbers of Far Eastern Original Equipment Manufacturers (OEMs).

Traditionally, the domestic kettle supply chain was characterised by a limited number of organisations, which were responsible for manufacturing kettle bodies, fitting controls and thermostats and distributing the final product directly to retailers in Western Europe.

Since the latter half of the 1990s, the availability of inexpensive kettles from Far Eastern OEMs meant that Western brands could no longer compete on price by manufacturing the products themselves. Therefore recent years have seen a trend in brands outsourcing products from the Far East that they can re-brand. The latter part of 2003 saw the last brand to manufacture kettles in the British Isles cease production, choosing to source their entire product portfolio from overseas.

The major logistical effect of the growth in the Far Eastern market is a substantial increase in variability of sales demand. This variability can be seen as a requirement for shorter lead times and a greater span in demand volume.

The nature of business in the Far East also differs significantly from the traditional Western approach. In the Far East, manufacturing is characterised by numerous companies utilising highly labour-intensive, albeit generic, methods. It is quite common for these companies also to shift their product output from one type of domestic appliance to another after a batch has finished, perhaps never resuming manufacture of the first item. For example, kettles may be made one month, toasters the next. Consequently, these manufacturers choose to work with large batch sizes in series as they win orders. This compares with the more stable Western European approach of manufacturing in smaller batches, producing lower volumes of different domestic appliances in parallel.

Not only has the supply chain evolved to include new Far Eastern OEMs but the growth in global markets has also resulted in traders and agents becoming involved in the buying and selling interactions between the various companies and players in the chain. In summation, the structure of small domestic appliance supply chain is now far more complex and dynamic than it was 10 years ago, and rarely stays the same from one day to the next.

9.3 The Just-in-Time (JIT) Paradigm

The dominant paradigm for addressing problems of responsiveness has been Just-in-Time (JIT). According to Harrison and Petty (2002), JIT can be described as "... *an approach that attempts to systematically eliminate waste*". JIT has been highly influential with many companies adopting it as a means of systematically improving manufacturing performance. The JIT perspective on the kind of planning problems being faced by Strix and other companies is that these are the result of long lead times. The JIT solution would be to attack the fundamental reasons for long lead times so that complex planning systems would not be necessary. If lead-time times are short and lot sizes small, then simple pull systems can be used as alternative to more sophisticated planning systems. There is no doubt that the JIT philosophy has been highly successful, but where capacity is geographically distributed, transhipment durations place a lower bound on the lead-times that can be achieved. In many companies, where JIT has been embedded into the organisational culture, the very existence of planning systems is seen as a sign of weakness. Companies such as Strix need to accept that the embracement of JIT philosophy in isolation is not sufficient in order to improve planning. It needs a combination of both process improvements and intelligent IT capabilities.

9.4 Literature Review

A review of the literature indicates that research relevant to planning in Dispersed Manufacturing Networks falls into two categories: SCM and marketing. Researchers in each field have identified different problems and causes, but are consistent in their general recommendations for future research. Nonetheless, the two categories provide a convenient structure for a review of the relevant literature.

9.4.1 Marketing Perspective

Researchers in the field of marketing have provided considerable insight into the changing nature of consumer markets, and consequently into the need that companies have for more detailed information and improved information systems. Traditionally, marketing managers and researchers act at a strategic level, concerned with the development of marketing strategies. The purpose of marketing strategy development is to establish, build, defend and maintain competitive advantage (McDonald, 1992). In the past, the routes to competitive advantage have typically been based upon strong brands, corporate image, effective advertising and, in some cases, price. These are the classic components of conventional marketing strategies.

As far back as the 1990s, Brady and Davis (1993), however, argued that the power of the brand, in both consumer and industrial markets, is in decline.

Market researchers have developed high levels of expertise in the analysis of market behaviours and so this could be a possible reason for the majority of marketing research literature appearing to focus on the changes in market perceptions. However, research into the nature of global markets and the interactions between companies with cultural and geographical differences is becoming more prevalent.

Recent literature shows how the field of marketing is broadening its horizons from simply facilitating transactions between companies and individuals to understanding market behaviours, structures and trends. Research, however, is still largely customer-focused and aimed at a strategic level.

Baker (2003) discusses the typical focus of current research. "*For whatever reason, the customer is seeking more than 'brand value', as it is sometimes called, and is looking increasingly for value in a much wider sense. This value might either be derived through the delivery of benefits in performance terms and/or in the form of a reduction in the customer's costs.*" Increasingly, customer relationship management (CRM) is being viewed as a strategic approach that can help realize improved customer value. In the new paradigm of marketing, the emphasis changes from brand value to customer value. Essentially this means that the supplying organisation must focus its efforts upon delivering an *offer* or *package* that will impact customers' perception of the value they derive through ownership of that offer (Baker, 2003).

Whatever the reasons behind the marketing problems faced by today's companies cited in the literature, there is a general consensus amongst researchers and practitioners alike that there is a need for managers to have more, higher quality external information regarding markets, their structures and behaviours. Marketing researchers consider external information to be of strategic importance, since strategic decisions are primarily long term with a balance towards external focus, whereas operational decisions are primarily short term and have an internal focus (Xu and Kaye, 1995).

Companies with superior information enjoy a competitive advantage. The company can choose markets better, develop better offerings and execute better marketing planning (Kotler, 2000). After reviewing the literature and based on practical experience, the authors would go further than this, suggesting that companies could also utilise better information resources to focus attention on more desirable transactions between customers and suppliers. Buttery and Tamaschke (1996) state "*A good marketing information system can make decision making more efficient and effective*". A marketing information system can be used to help create a competitive advantage and can even substitute for expensive assets. In recent years, the use of computer-based information systems in developing marketing strategy has gained attention from academic researchers. Decision Support Systems (DSSs) have been developed to assist with the formulation of marketing strategy through the use of quantitative models and analytical techniques (Li *et al.*, 1999).

Given the value attributed in the literature to marketing information and marketing information systems, it might be expected this would be reflected in industrial practice. However, this is does not appear to be the case. Current decision support systems for marketing, in most instances, exist in a rudimentary stage of development (Li *et al.*, 1999). For example, in a survey of 100 chief executives, less than 10% of companies (in the service and other sectors) indicated that they had a fully-integrated marketing information system, less than one-quarter of respondents were satisfied with access to internal information and two thirds were either dissatisfied or very dissatisfied with access to external information (Buttery and Tamaschke, 1996)

This is surprising as in the early 1990s, Jones and Arnett (1993) pointed out that "*The effectiveness of Information Systems (IS) design and development ... are key issues for the 1990s*". As we are now well into the 21st century, why might development have been so slow? Many Small and Medium Enterprises (SMEs) lack a dedicated marketing department to conduct research and even larger companies typically limit work to routine forecasting based on sales analysis and occasional surveys (Kotler, 2000). In the case of Strix, with around 1000 employees worldwide, it would be impossible to dedicate resources to detailed market research.

9.4.2 SCM Perspective

Globalisation has been a key theme for SCM researchers in recent years. Christopher (1992) classifies companies that need to organize globally into three groups:

- Firstly, there are the commodity companies where the task is that of moving bulk raw materials from countries with surplus natural resources to those with either the markets to consume them, the labour to process them or both.
- Secondly, there are the companies who are taking advantage of low regional labour costs to maximise profitability on labour-intensive manufacturing.
- Finally, there are the companies who have chosen to concentrate their investment in R&D and manufacturing, focusing each of their sites on specific product-technology combinations.

Strix Ltd, like many western SMEs, would fall very naturally in the second group with the added objective of gaining access to the fast growing far eastern market.

The logic of the global company is clear: it seeks to grow its business by extending its markets whilst at the same time seeking cost reduction through scale economies in purchasing and production and through focused manufacturing and/or assembly operations. However, whilst the logic of globalisation is strong, it must be recognised that it also presents certain challenges. Firstly, world markets are not homogeneous, so there is still a requirement for local variation in many product categories. Secondly, unless there is a high level of co-ordination, the complex logistics of managing global supply chains may result in higher costs. (Christopher, 1992)

There is a danger that some global companies in their search for cost advantage may take too narrow a view of cost and only see the cost reduction that may be achieved by focusing on production. In reality it is a total cost trade-off where the costs of longer supply pipelines may outweigh the production cost saving.

Increasing numbers of SMEs are becoming active in international business. This phenomenon is being facilitated by numerous trends in the macro-environment of the firm. However, SMEs are typically quite resource-constrained relative to their larger, traditional rivals. Many SMEs fail to bridge the gap between initial exporting efforts and ultimate success abroad (Knight and Yaprak, 2000).

The trend towards global manufacturing is highlighting companies who place the organisation of logistics and the supply chain as a critical success factor. The complexity of the logistics task appears to be increasing exponentially, influenced by such factors as the increasing range of products, shorter product lifecycles, marketplace growth and the number of supply/market channels. SME companies frequently lack the appropriate structures or organisational format to accommodate the uncertainty dynamic business environments place on them (Xu and Kaye, 1995). As such many researchers in the field of SCM have focused their efforts on supply chain integration and collaboration between companies.

The increasing complexity of supply chains and focus on supply chain integration may also be responsible for a shift in terminology within the SCM arena. Traditionally, a supply chain involves only a single enterprise with multiple facilities and distribution centres. In recent years, however, the scope of SCM has evolved to cross the enterprise boundaries (Gan *et al.*, 2000). Indeed "Supply Network" may now be a better name to represent real life supply chains (Jain *et al.*, 2001).

Gupta *et al.* (2001) typify much of the recent research in SCM and, as with marketing, identifies that improved information resources will be critical for organisations in the near future. They state "*To maximize value to the customer, the entire supply chain must be 'optimised.' A supply chain will continually lose market share if each component of the supply chain is not involved in continuous improvement.*"

In fact, the management of global logistics has in reality become the management of information flows. The information system is the mechanism whereby the complex flows of materials, parts, subassemblies and finished products can be coordinated to achieve cost-effective service. Any organisation with aspirations to global leadership is dependent upon the visibility it can gain of materials flows, inventories and demand throughout the pipeline. Without the ability to see down the pipeline into end-user markets, to read actual demand and subsequently to manage replenishment in virtual real-time, the system is doomed to depend upon inventory. (Christopher, 1992)

Intuitively, one can argue that information sharing could or can substantially improve the overall supply network performance, however, unlike marketing, SCM researchers tend to focus on information requirements of tactical and operational levels. Sharing information such as demand, sales orders, inventory status and order fulfilment status could help companies to reduce inventory cost, shorten time-to-market, and improve decision making along the total supply chain. Consequently, customer service could be improved. Thus, information sharing boosts the efficiency and performance of a supply chain (Ball *et al.*, 2002).

Within the supply chain several different types of information can be exchanged. Young (2000) claims that "*the better a group of organizations becomes at passing information throughout the supply chain, the more efficient that supply chain becomes*". Recently, two kinds of enterprise systems have been widely adopted in industry to facilitate information sharing and understanding of supply networks: Enterprise Resource Planning (ERP) systems and Supply Chain Management Systems (SCMS). ERP systems are able to provide an integrated transaction processing infrastructure for an organisation, which enhances organisational performance by reducing information inconsistency and by improving transaction-processing efficiency. SCMS, on the other hand, are aimed at providing a higher level of business planning and decision support related to activities that involve the coordination and execution of multi-organisation wide production and distribution processes (Ball *et al.*, 2002). However, despite these recent advancements in information technology, effective information sharing between members of a supply chain is still limited. Moreover, there has been much criticism among academics and practitioners with respect to the implementation of ERP systems. According to recent research by one of the authors (Labib, 2008) "*most existing off-the-shelf software packages, especially CMMSs and enterprise resource planning (ERP) systems, tend to be 'black holes'*". The term "black hole" is used to indicate that systems require a large amount of data to support their operation, yet provide relatively little assistance in terms of decision support. Companies often expend considerable effort in servicing such systems yet find that that the only support for decision making is the data that they themselves entered, albeit sorted and consolidated. In summary, Labib (2008) states, "*In short, companies tend to spend a vast amount of capital in acquisition of off-the-shelf systems for data collection, but their added value to the business is questionable*".

9.5 A Survey of UK Companies

Whilst the literature in this area is rich, there is a shortage of quantitative data on the scope and scale of capacity dispersion. Because of this, it is difficult to be sure that the problems being faced by Strix are unusual. To overcome this deficiency it was decided to undertake a survey of UK companies. It was decided that the questionnaire should be brief in line with its restricted aims. Moreover, it was felt to be important that a high response rate would be beneficial and a brief questionnaire would be helpful in achieving this end. A total of 460 companies were contacted by email and invited to complete an on-line questionnaire. The overall response rate was 45%. An analysis of the questions asked is presented below (see the Appendix for details of the questionnaire).

9.5.1 Company Size

Figure 9.3 shows that the size distribution of SME companies who responded to the questionnaire approximates a binomial distribution, the majority of respondents having between 100–200 employees.

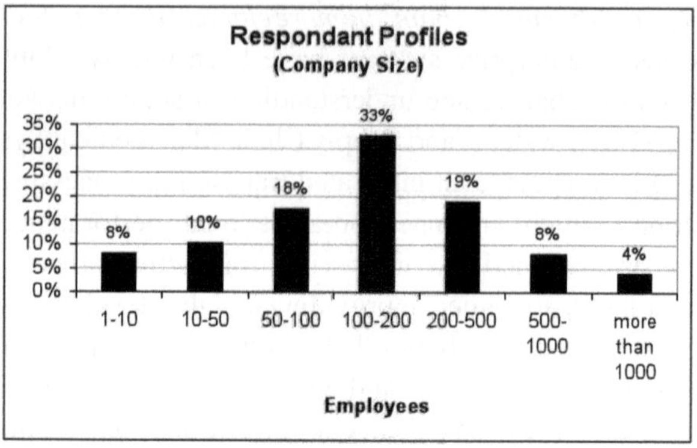

Figure 9.3. *Respondent profiles by company size*

9.5.2 Positions in the Supply Chain

Figure 9.4 shows how the majority of respondents were either 2nd or 3rd tier suppliers. The total percentage is greater than 100, but this is because some respondents declared that they act at more than one level in the supply chain. Another point worth mentioning is a result of analysing the comments made by the respondents. Some respondents stated that they were either unable to answer where they lay in the supply chain for a few reasons:

- The diagram attached to the questionnaire showed a "Retailer" level. Some respondents reported that they did not believe retailers played a part in the supply chains of which they were a member.
- A few respondents mentioned that they did not know how many tiers there were between themselves and the end consumer. This information is valuable as another deduction can be made. Those companies that do not know how many tiers there were between them and the end consumer are unlikely to engage in any supply chain intelligence gathering.
- Some respondents showed a greater awareness of the supply networks in which they operate, declaring that the nature of their supply chains were too complex to be able to place themselves at a particular level/tier.

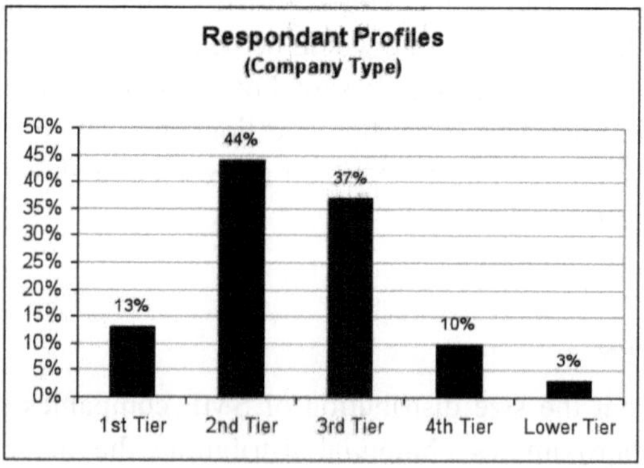

Figure 9.4. *Respondent profile by supply chain position*

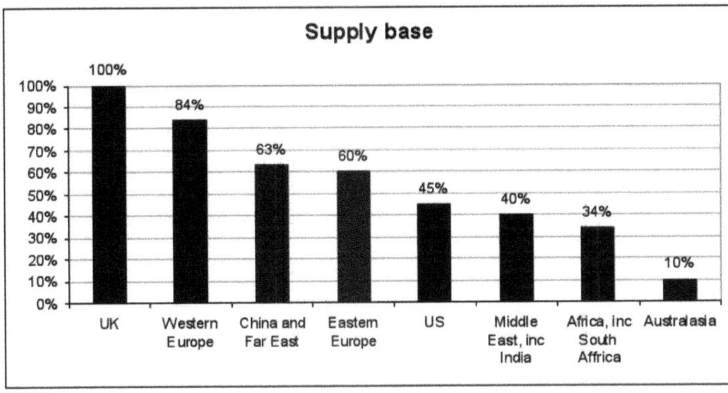

Figure 9.5. *Supply base profiles for respondents*

Figure 9.5 shows that all of the respondents have at least one supplier within the UK and that while Western Europe is still an important source of supply, China and the Far East is significant.

Figure 9.6 shows China and the Far East are not only an important source of supply, but are also an important customer base. It was not surprising that China is an important customer base, but just how important was surprising. The conclusion drawn by the authors is that the large population and rapidly expanding economy is making the Far East an ever more attractive market. When the results are compared with Figure 9.5 and the case study company another deduction can be drawn. It would appear that Strix are not alone and indeed many 2nd and 3rd tier suppliers are selling large quantities of products to 1st tier suppliers in the Far East. 1st tier suppliers are typically characterised by large-scale assembly operations.

9.5.3 Companies Choosing to Expand across Geographical Boundaries

A key aim of the questionnaire was to investigate the degree to which UK companies are choosing to locate manufacturing, logistical or service capabilities in the Far East or Eastern Europe. 27% of the respondents stated that they had recently expanded

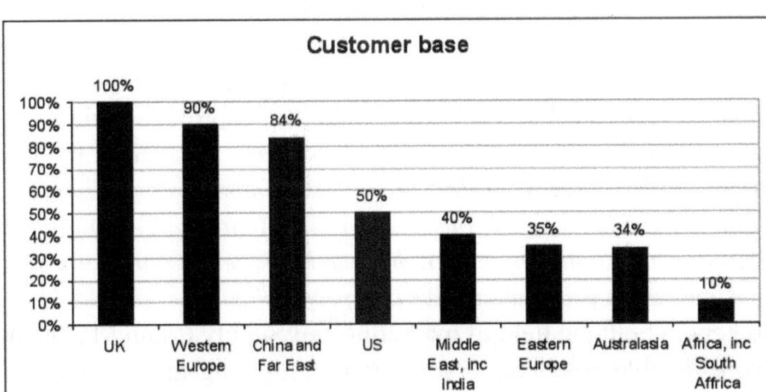

Figure 9.6. *Customer base profiles for respondents*

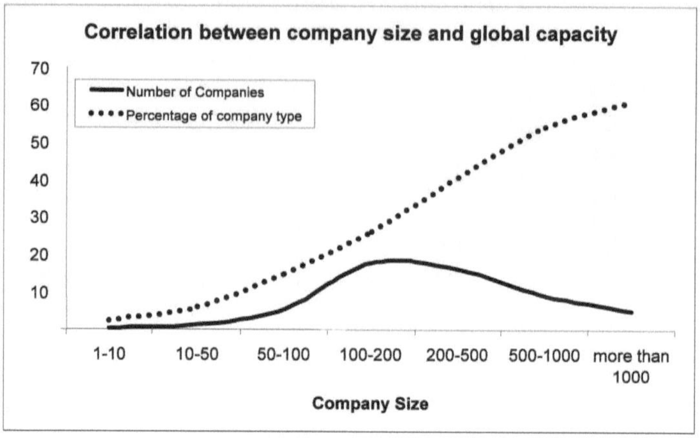

Figure 9.7. *Correlation between company size and global capacity*

operations in other countries. Presented in such simplistic terms, the information is of limited value, however, when a correlation is made between the company types and sizes (see Figure 9.7) then more sophisticated inferences can be drawn.

Figure 9.7 shows that, of those companies who have expanded their capabilities, a greater percentage of larger companies have expanded their capabilities across geographical boundaries than smaller companies. It could be that larger companies have greater resources available and so are more able to expand.

Figure 9.8 also shows that, of those companies who have expanded their capabilities, the profile of company types is more evenly distributed. 2nd and 3rd tier suppliers constitute the majority of companies who have expanded their operations across geographical boundaries whilst only a small percentage of 1st tier suppliers have. It may be that 1st tier suppliers are facing tough competition from less expensive imports from the Far East and Eastern Europe.

9.6 Conclusions

The globalisation of industrial activity has become a major issue in business. Articles in the business press, seminars and academic symposia have all focused upon on this emerging global trend. This chapter has described the problems faced by one company, Strix Ltd. The survey presented above provides evidence that Strix is not unique. It has shown that these issues have been recognised by a number of researchers. Indeed Xu and Kaye (1995) state "*All companies are navigating into an uncertain future environment. The business environment is increasingly volatile and turbulent, within which senior managers have to manage. Organizations must adapt to their environments in order to survive and prosper*".

Researchers in both the field of marketing and Supply Chain Management have expended considerable resources and effort investigating the problems caused by globalisation of organisations and attempting to find solutions to specific problems. The competitive pressures and challenges that have led to this upsurge of research

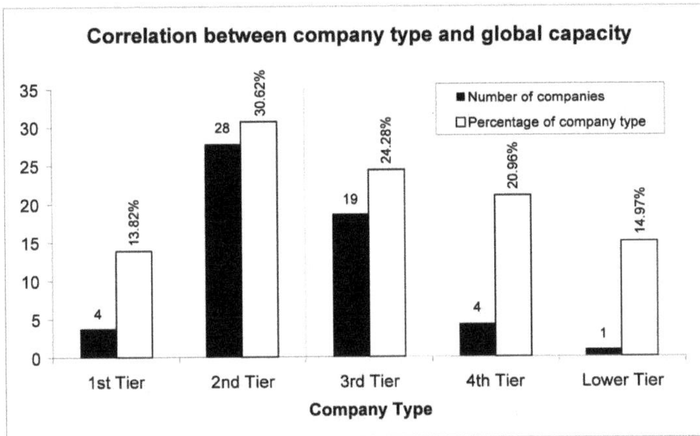

Figure 9.8. *Correlation between company size and global capacity*

interest have been well documented. What is less well understood are the implications of globalisation for operations management in general and specifically for logistics management.

It is not surprising to find that both marketing and SCM experts conclude that further research needs to be conducted in their respective fields. There is, however, a common consensus within the literature that:

- There is a real need for more, higher quality information to be gathered about the nature of supply chains and markets.
- Current information systems present within organizations are lacking or have failed to bring about a significant improvement. Labib (2008) provides a more detailed account of the state of such systems in the area of ERP and maintenance systems.
- Companies should use information technology to develop intelligence in systems, in order to scan and monitor environment changes.

The very fact that marketing managers are calling for more external information reflects a failure of existing systems in supporting managers' information requirements (Xu and Kaye, 1995).

In summary, many manufacturers are operating across national boundaries and this has led to increased complexity in managing the organisation. A key issue is the increased lead-time that results from transhipment. This issue is not easily addressed by the tools of JIT, which have been applied routinely over the last twenty years. This means companies will need to place greater emphasis on planning systems and the information systems that support them. This greater emphasis needs to be focused on provision of decision support, which can be achieved by embracing a transformation from data collection to decision analysis. Operational Research techniques such as multiple criteria for prioritisation and selection, as well as optimisation techniques will ensure that such systems are both efficient and effective.

Appendix

The questions asked in the survey were all multiple-choice and are as follows:

1. How many employees are in your company, including directors?
   a. 1–10
   b. 10–50
   c. 50–100
   d. 100–200
   e. 200–500
   f. 500–1000
   g. More than 100
2. Based upon the diagram shown below, at which tier(s) would you say your company is positioned?

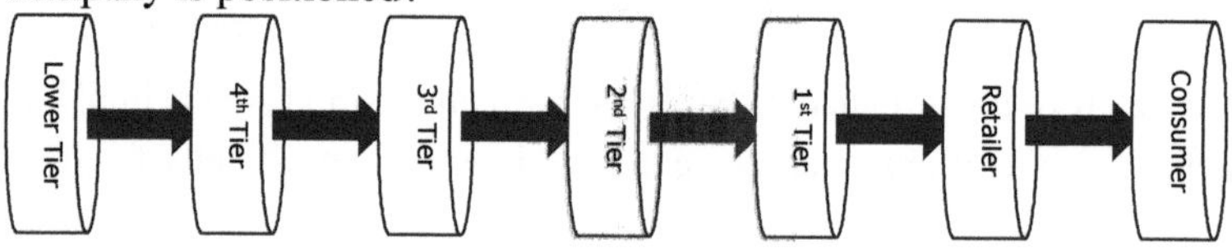

3. Please indicate which regions your company sources materials, components and or services from:
   a. UK
   b. US
   c. Western Europe
   d. Eastern Europe
   e. China and Far East
   f. Middle East, incl. India
   g. Africa, inc South Africa
   h. Australasia
4. Please indicate which regions your company's direct customers are in
   a. UK
   b. US
   c. Western Europe
   d. Eastern Europe
   e. China and Far East
   f. Middle East, incl. India
   g. Africa, inc South Africa
   h. Australasia
5. Within the last 10 years, has your company opened any manufacturing, logistical or service capabilities within the developing regions of the Far East, Eastern Europe or Middle East? Yes/No

Finally an area where the respondent could add any comments about the questionnaire was provided.

References

Argyris, C. and Schön, D. (1989), "Participatory Action Research and Action Science Compared: A Commentary", American Behavioral Scientist, Vol. 32, No. 5, pp. 612–623.

Azevedo, A.L. and Sousa, J.P. (2000), "A component-based approach to support order planning in a distributed manufacturing enterprise", Journal of Materials Processing Technology, Vol. 107, No. 1–3, pp. 431–438.

Baker, M. J. (2003), 'The Marketing Book', Butterworth Heinemann, Oxford.

Ball, M.O., Ma, M., Raschid, L. and Zhao, Z. (2002), 'Supply Chain Infrastructures: System Integration and Information Sharing', Sigmod Record, Vol. 31, No. 1, pp. 61–66.

Brady, J. and Davis, J. (1993), 'Marketing Mid-life Crisis', McKinsey Quarterly, No. 2, pp. 17–28.

Buttery, A. and Tamaschke, R. (1996), 'The use and development of marketing information systems in Queensland, Australia', Marketing Intelligence & Planning, Vol. 14, No. 3, pp. 29–35.

Christopher, M. (1992), 'Logistics and Supply Chain Management Strategies for reducing Costs and Improving Services', Managing the Global Pipeline', Financial Times Pitman, London.

DTI – Department of Trade and Industry (2002), The Government's Manufacturing Strategy, Department of Trade and Industry, available at: www.dti.gov.uk/manufacturing.

Duffie, N.A. and Prabhu, V.V. (1996). "Heterarchical control of highly distributed manufacturing systems", International Journal of Computer Integrated Manufacturing, Vol. 9, No. 4, pp. 270–281.

Frederix, F. (2001). "An extended enterprise planning methodology for the discrete manufacturing industry", European Journal of Operational Research, Vol. 129, No. 2, pp. 317–325.

Gan, B.P., Liu, L., Jain, S., Turner, S.J., Cai, W. and Hsu, W. (2000), 'Distributed Supply Chain Simulation Across Enterprise Boundaries', Proceedings of the 2000 Winter Simulation Conference, pp. 1245–1251.

Gupta, A., Whitman, L. and Agarwal, R.K. (2001), 'Supply Chain Agent Decision and System (SCADAS)', Proceedings of the 2001 Winter Simulation Conference.

Harrison, D.K. and Petty, D.J. (2002), "Systems for Planning & Control in Manufacturing", Newnes Publishing, London.

Jain, S., Workman, R.W., Collins, L.M., Ervin, E.C. and Lathrop, A.P. (2001), 'Development of a High-Level Supply Chain Simulation Model', Proceedings of the 2001 Winter Simulation Conference, pp. 1129–1137.

Jones, M.C. and Arnet, K.P., (1993), 'Current Practices in Information Management Systems', Information and Management, Vol. 24, No. 2, pp. 61–69.

Harland, C.M. (1996). "Supply Chain Management: Relationships, Chains and Networks", British Journal of Management, Vol. 7, Special Issue, pp. S63–80.

Knight, G.A. and Yaprak , A. (2000), 'Emerging patterns of international business: Implications for public policy', Advances in International Marketing, Vol. 10, pp. 385–397

Karlsson, C. (2003). "The development of industrial networks: Challenges to operations management in an extraprise", International Journal of Operations & Production Management, Vol. 23, No. 1, pp. 44–61.

Kotler, P. (2000), 'Marketing Management, Millennium Edition, Part II", Prentice Hall, Englewood Cliffs, NJ.

Labib, A.W. (2008), Computerised Maintenance Management Systems, in "Complex Systems Maintenance Handbook", Edited by: K.A.H. Kobbacy and D.N.P. Murthy, Springer, Berlin.

Lambert, D.M. and Cooper M.C. (2000), 'Issues in Supply Chain Management – Don't Automate, Obliterate', Industrial Marketing Management, Vol. 29, No. 1, pp.65–83.

Li, S., Duan, Y. Kinman, R. and Edwards, J.S. (1999), 'A framework for a hybrid intelligent system in support of marketing strategy development', Marketing Intelligence & Planning, Vol. 17, No. 2, pp. 70–77.

Locomonitor (2005), Foreign direct investment by country – World economic region, available at: http://www.locomonitor.com/index.cfm.

Mason-Jones, R., Naylor, B. and Towill, D.R. (2000). "Engineering the leagile supply chain", International Journal of Agile Management Systems, Vol. 2, No. 1, pp. 54–61.

Maturana, F.P. and Norrie, D.H. (1996). "Multi-agent Mediator architecture for distributed manufacturing", Journal of Intelligent Manufacturing, Vol. 7, No. 4, pp. 257–270.

McDonald, M. (1992), 'Strategic Marketing Planning; A state-of-the-art review.', Marketing Intelligence & Planning, Vol. 10, No. 4, pp. 4–22.

O'Neill, H. and Sackett, P. (1994). "The Extended Manufacturing Enterprise Paradigm", Management Decision, Vol. 32, No. 8, pp. 42–49.

Preston S. (2004), "Lost in migration: offshore need not mean outsourced", Strategy & Leadership, Vol. 32, No. 4, pp. 32 – 36.

Shen, W. (2002). "Distributed Manufacturing Scheduling Using Intelligent Agents", IEEE Intelligent Systems, Vol. 17, No. 1, pp. 88–94.

Xu, X. and Kaye, G.R. (1995), 'Building market intelligence systems for environment scanning', Logistics Information Management, Vol. 8, No. 1, pp. 22–29.

Young R.R. (2000), 'Managing Residual Disposition: Achieving economy, environmental responsibility; and competitive advantage using the supply chain framework', Journal of Supply Chain Management, Vol. 36, No. 1, pp. 57–66.

10 Managing Product Variety in Multinational Corporation Supply Chains: A Simulation Study Investigating Flow Time

Mahendrawathi Er
Institut Teknologi Sepuluh Nopember

Bart MacCarthy
University of Nottingham

Abstract Today's business environment is characterised by globally dispersed supply and manufacturing networks. In addition, the level of product variety continues to increase in almost all sectors. Managing product variety in the context of international operations has received very little attention to date in the research literature. Here we present findings from a simulation study investigating the impact of product variety, supply lead-time and demand uncertainty on production flow time within a multinational corporation (MNC) context. A generic simulation model is used to investigate the impact of increasing product variety in combination with supply lead-time and demand uncertainty in an international manufacturing setting. The simulation focuses on the upstream activities of forecasting, inbound supply and manufacturing. The structure and logic of the simulation model are based on insights gained from an empirical study of a real MNC supply network.

The study shows that increasing the level of product variety has a detrimental impact on flow time. In the presence of supply lead-time and demand uncertainty, high levels of variety result in much longer flow times relative to more stable conditions. The impact is greatest when variety involves critical materials that are required early in the production process and that entail long set-up times. The value of the study is that it quantifies the potential relative impacts of different factors on production flow time in an MNC context. Implications of the study for managing international operations are discussed, particularly with respect to the in-bound supply of materials. The potential value of postponement strategies and the need in some cases for fundamental product and process redesign to mitigate the negative impacts of variety are noted. The study could be extended to incorporate downstream distribution, logistics and transport activities, multiple manufacturing sites and multiple potential supply routes.

Keywords Flow time, Multinational Corporations, Product variety, Simulation, Supply Chain Management, Uncertainty

10.1 Introduction

More and more companies are involved in international supply chains in various ways, ranging from simple import and export activities to the development of subsidiaries in foreign countries. International issues are a common characteristic of today's supply chains (Akkermans, 1999; Wisner *et al.*, 2004). In combination with the internationalisation of supply chains, product variety has been growing in almost all sectors (Cox and Alm, 1998; Bils and Klenow, 2001).

In any operational context, high levels of product variety may result in set-up and changeover delays and may require more complex procurement and demand management approaches (Fisher *et al.*, 1994; Randall and Ulrich, 2001). When the supply network is dispersed around the world, the challenge is even greater with potentially longer and more uncertain delivery times, as well as inflexibilities in planning, procurement and ordering systems (Levy, 1995; Lowson, 2001). Managing product variety in an international setting is a challenging and important task. The combined literature on international business, supply chain management and product variety is extensive. However, the impact of product variety in relation to factors usually found in the context of international operations has not been addressed.

This chapter investigates the issue of product variety in typical Multinational Corporation (MNC) supply chains. The main objective is to examine the impact of variety in conjunction with other issues of importance in the international context, particularly demand uncertainty and supply lead-time uncertainty. The approach used is to investigate through simulation the performance of an MNC supply chain in order to identify key issues and to indicate potential strategies to address them. The focus is on the upstream activities of production planning, inbound supply and manufacturing rather than downstream activities of distribution to markets and customers.

In the next section, relevant literature is reviewed. Empirical fieldwork that provided the basis for the development of a simulation model is then described briefly in Section 10.3. This is followed by a detailed description of the simulation work, which includes the simulation environment, the assumptions made, simulation experimentation issues and performance measures. Results from the simulation experiments are discussed in Section 10.5. Section 10.6 describes the implications of the findings for practice. Brief concluding remarks and directions for further research are provided in the final section.

10.2 Literature Review

This work builds upon several streams of literature including international operations, supply chain management, international supply chain management and product variety.

Companies are driven to develop dispersed manufacturing networks for many reasons, including achieving lower costs, accessing new markets, seeking to acquire strategic assets such as a skilled workforce or special technologies (Ferdows, 1989; MacCarthy and Atthirawong, 2003). Multinational Corporations (MNC) are an important part of today's business environment. Ghoshal and Bartlett (1990) note that an MNC consists of a group of geographically dispersed and goal-disparate organisations that includes its headquarters and the different national subsidiaries.

According to Porter (1986), the distinctive issues in international, as opposed to purely domestic operations, can be summarised in two dimensions – configuration and co-ordination. Configuration is concerned with location and structure – where activities are performed and in how many places. Co-ordination is concerned with the linkages between different activities of companies operating internationally, such as planning and scheduling (Oliff *et al.*, 1989). Effective supply chain configuration and co-ordination strategies may enable a MNC to achieve strategic advantages over its competitors in terms of costs and market responsiveness.

One of the key issues found in the management of a MNC is the choice between global or local sourcing for raw materials, parts, components or products. The availability of low-priced commodities in the world market may force MNCs to source globally instead of domestically (McGrath and Bequillard, 1989). Some MNCs decide to source from trusted suppliers, despite their location, in order to procure high quality materials. However, sourcing from global suppliers located far from manufacturing units may lead to long and highly uncertain delivery times (Levy, 1995). Based on his study on North American and UK retailers, Lowson (2001) indicates that domestic suppliers are more flexible and responsive to accommodate volume and mix changes compared to international suppliers.

Product variety has received attention in a separate stream of literature (Ramdas, 2002). Several authors investigate the impact of product variety on manufacturing performance. Anderson's (1995) study in three textile weaving plants of a single firm indicates that manufacturing overhead costs increase with the number and severity of set-ups. MacDuffie *et al.* (1996) studied 70 automotive assembly plants world wide and found varying impacts of different types of product variety on labour productivity. An empirical analysis by Fisher and Ittner (1999), also in the automotive industry, indicates that in mixed model assembly operations, variability in product mix may be a more important indicator of the effects of variety than measures such as the number of products or number of parts, commonly used in research studies and activity-based costing systems.

More recently, some studies have investigated the issue of product variety in the context of supply chains. Randall and Ulrich (2001) suggest that there is a coherent way to match product variety with supply chain structure and that firms that correctly match the types of variety offered with the supply chain structure perform better compared to those that fail to do so. Thonemann and Bradley (2001) presented a mathematical model to analyse the effect of product variety on the performance of a supply chain with a single manufacturer and multiple retailers. They demonstrate that disregarding the effect of product variety on lead-time can result in poor decisions with companies offering a level of product variety that is greater than optimal.

Product variety may result from differences in materials and/or production processes at various stages of the supply chain. Several authors have noted the potential effect of product variety on supply systems (MacDuffie *et al.*, 1996; Milgate, 2001). More variety potentially adds complexity to the configuration and co-ordination of supply networks (Milgate, 2001).

High product variety also creates uncertainty in demand (Randall and Ulrich, 2001). Fisher *et al.* (1994) argue that having a wider range of product variants means that it is more difficult to predict demand at the product level. In the presence of demand uncertainty, it is difficult to precisely match supply with demand. Mismatches cause disruptions to production, particularly when demand exceeds supply. Variety also generates costs associated with holding inventory and product markdowns when supply exceeds demand and the costs of lost sales when demand exceeds supply (Randall and Ulrich, 2001).

The challenge of managing product variety is amplified further when product manufacturing is dispersed across national boundaries. Geographical distance between elements in international supply networks may be associated with long lead-times and greater delivery uncertainty (Levy, 1995). As a consequence, some activities, particularly procurement of materials, must be conducted some time in advance of real demand signals. This exposes the supply network to greater risks of a mismatch between the supply of materials procured based on forecasts and the actual demand that ensues. It may also limit the supply network's ability to respond to demand changes in the marketplace.

Simulation is a tool commonly used to address a range of issues in operations management and has gained popularity in supply chain analysis due to its strength in predicting system variation and interdependencies (Wyland *et al.*, 2000). It has been used to investigate the impact of demand and supply uncertainty on supply chain performance. In addition, it has also been used to investigate the impact of product variety on manufacturing performance. Several studies that have used simulation to address these issues are summarised in Table 10.1.

The review of the various streams of literature above indicates that investigating product variety in a supply chain operating internationally is a timely issue that needs to be pursued. The issues relating to product variety and international operations have not been addressed simultaneously in the literature to date. This study aims to fill part of this gap by addressing the following research question: *What is the impact of product variety on flow time in an MNC supply chain operating internationally in the presence of supply and demand uncertainty?* To address the research question, both an empirical and a simulation study have been conducted. The purpose of the empirical study was to provide insights on: 1) the characteristics of companies operating internationally in terms of planning and control and 2) the issues facing these companies with respect to product variety, supply and demand uncertainty. A simulation study was then conducted to investigate more generally the impact of product variety in the context of an enterprise operating internationally. The results from the simulation study are the focus of this chapter. The empirical study is described very briefly first.

10.3 Empirical Study

In the initial part of this work, we conducted a fieldwork study involving eleven different manufacturing companies in Indonesia and one company in the UK belonging to international supply networks (Er, 2004; Er and MacCarthy, 2002). Semi-structured interviews were conducted to gather information on their network characteristics and relevant issues in managing international supply chains. Four

Table 10.1. *Examples of simulation studies in the literature*

Literature	Focus of the study	Description of study
Fisher and Ittner (1999)	Product variety	Indicates that the use of direct labour slack is an optimal response to increased product variety in terms of option variability
Levy (1995)	Demand and supply uncertainty	Demand instability raises the proportion of unfulfilled demand for different international supply chain configurations. Disruption in supplier deliveries reduced demand fulfilment and increased inventory levels
Towill (1996)	Supply chain dynamic behaviour	Used System Dynamics in supply chain redesign to generate added insights of a system's dynamic behaviour and the underlying causal relationships
Ovalle and Marques (2003)	e-collaboration tools	Developed a System Dynamics model to assess the impact of using e-collaboration tools on the supply chain performance
Spengler and Schroter (2003)	Integrated production and recovery system	Used System Dynamics to model an integrated production and recovery system for supplying spare parts
Persson and Olhager (2002)	Supply chain design	Developed a discrete event simulation model to evaluate alternative supply chain designs
Al-Zubaidi and Tyler (2004)	Retailing and supply procedures	Developed a discrete simulation model to investigate the effects of improved retailing and supply procedures on financial and other performance measures
Tiger and Simpson (2003)	Material flow	Developed a discrete-event simulation to assist a Multinational Company in understanding the impact of material flow from the US to the Asia Pacific region

of the companies provided the most relevant insights specifically for MNC supply chains. The four were subsidiaries of MNCs producing leather shoes, ladies underwear and lightbulbs (two companies). Analysis of the empirical data from the four companies highlighted the following common characteristics:

1. **Configuration:** the major elements in the four MNC supply chains were suppliers, a manufacturer, sales office, wholesalers and corporate headquarters. Sales offices, wholesalers or distributors and in some cases corporate headquarters were in charge of demand management activities. We refer to these elements as 'internal customers' that transform demand information from end customers into production demand for the manufacturer.
2. **Co-ordination:** production is triggered by demand from the internal customer. Planning and forecasting is typically done with centralised control from headquarters.
3. **Products:** products are not technologically complex, typically they are consumer products.
4. **Manufacturing process:** manufacturing can be classified as primarily discrete processing. In general, the manufacturing processes observed in the four cases

can be classified into three sequential stages of production. The first stage is typically associated with the main material preparation involving fabrication or manufacturing operations on main (raw) materials. These are followed by assembly processes. The final stage typically consists of finishing and packaging of finished products. Manufacturing of products is typically carried out in batches.

5. **Product variety:** variety is determined by the use of different types of materials at different stages of the production process.

The study also highlighted the problems of supply and demand uncertainty as major challenges in international supply chain management. Each company procures materials from a combination of global and local suppliers and faces supplier lead-time uncertainty. It was clear that global sourcing entailed longer lead-times and higher delivery uncertainty compared to local sourcing. Each manufacturing company also faces demand uncertainty as production requirements and forecast information from their internal customers are updated periodically to reflect the most recent conditions in the marketplace. We refer to this periodic updating of demand and forecasts as a 'rolling forecasting system'.

Findings from the empirical study have provided important qualitative insights on how MNCs manage product variety in international supply chains. However, such a study can provide only limited information on the magnitude of impact of product variety and related factors on overall performance. Any empirical study can give only 'point samples' of the complete 'operational space'. A generic simulation model allows us the opportunity to explore the range of potential behaviour and performance that may be observed or that may be possible. Using understanding and evidence from an empirical study, such a generic simulation study can be undertaken in an informed way. Thus, a simulation study was conducted to give greater insights. The characteristics of the MNC supply chain described above have been used as a platform to develop a generic simulation model for experimentation.

10.4 Simulation Study

Insights and information gained from the case companies on configuration have been used in developing the simulation model structure. Knowledge gained on co-ordination has provided the 'logic' for the information and material flows. Important and difficult problems facing the case companies, including demand uncertainty, supply lead-time uncertainty and increasing levels of product variety, are treated as factors to be investigated in the simulation. In particular, the simulation environment reflects the international MNC context in two ways: (1) by adopting centralised planning and forecasting resulting in long and inflexible planning lead-times, particularly to ensure materials supply, and (2) in the selection of parameter values and ranges used in experimentation.

The simulation model uses a discrete-event approach and has been developed using General Purpose System Simulator (GPSS) World™ (Minuteman Software, 2005). Full technical details of the model implementation are provided in Er (2004).

10.4.1 Simulation Environment

The simulation environment is illustrated in Figure 10.1. It captures a three-stage Multinational Corporation supply network consisting of 1) suppliers, 2) a manufacturer and 3) internal customers. At the beginning of a planning period the internal customer generates production demand and forecast information for the manufacturer. This information is updated from one period to the next reflecting a rolling forecast system, enabling *demand uncertainty* to be captured.

The manufacturer uses the information from internal customers to plan its production and procure materials. Figure 10.1 shows that the manufacturing process is divided into three typical stages – process 1, process 2, and process 3. Production of a specific product variant requires different types of materials, classified here as:

1. Main material (MM) representing raw material fabricated in process 1.
2. Unique material (UM) representing auxiliary materials, parts or sub-assemblies, assembled in process 2.
3. Packaging material (PM) representing packaging and general finishing items used in process 3.

Each type of material has a number of options designated by different numbers, e.g. main material 1, unique material 2, etc. The system produces different product variants through different options of main material, unique material and packaging material in manufacturing processes. Processing different materials at different stages of manufacturing requires set-up activities.

Each type of material can be obtained from a *local supplier* located close to the manufacturer or a *global supplier* located in a different country. The length of time required by a supplier to deliver the material is subject to uncertainty, referred to here as *supply lead-time uncertainty*. Procuring materials from global suppliers requires a relatively long lead-time compared to local suppliers. Thus, the manufacturer has to place orders further ahead of production if the materials are bought from global suppliers. Different types of material received from suppliers are stored in different inventory storage buffers as shown in Figure 10.1.

The production of an order is divided into *batches*. It is assumed that the size of transfer batch is equal to the process batch. We assume a relatively efficient and flexible plant and use the smallest feasible batch size. The size of a batch depends on the level of product variety. A more detailed description on the batch size is given in Subsection 10.4.6. The production sequence of batches is determined using a scheduling rule that minimises set-up time. Before a production schedule is executed, the availability of materials to produce each batch is checked. If sufficient materials for a batch are not available, the materials are not assigned to the batch. If materials are available, they are assigned to (pegged to) a specific batch and sent to the queue for the appropriate process.

At each processing stage, there is a queue to enter the process. As soon as process 1 is available, main materials are processed. The output of process 1 is then assembled with unique materials in process 2. As process 2 is completed, its output is packed and finished with packaging materials in process 3. At any stage of production, a batch is delayed if the correct materials are not available. Delayed batches wait in the corresponding queue for materials (e.g. queue wait MM) until

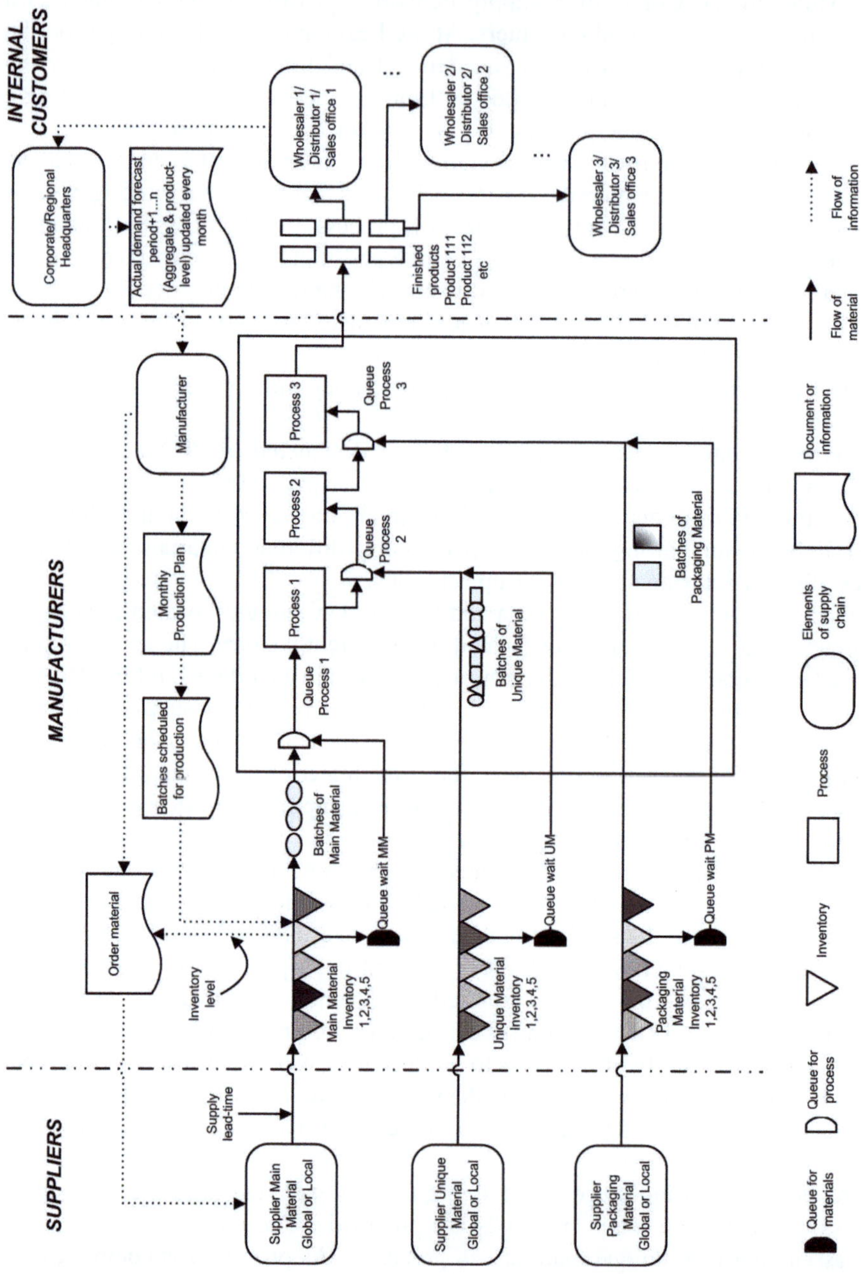

Figure 10.1. The simulation model

sufficient materials are available. Processing different material options at each stage of production introduces a delay due to set-up activities. Finally, finished products are allocated to satisfy the internal customer demand.

10.4.2 Assumptions

In developing the simulation model, several assumptions are made:

- The system works with an operational month of 20 working days. The capacity per month is assumed to be constant.
- No planned level of safety stock of materials and finished goods is assumed. This is an aspect highlighted for further work.
- The entire production process is conducted in the same plant and there is no time delay associated with transferring batches between production processes.
- A scheduling rule that minimises set-up time in production and ultimately minimises the average lead-time is applied. The rule first takes into account the time a batch is generated to ensure that batches generated earlier are given priority and processed first. Then, the rule ensures that batches with the same materials are done sequentially in order to reduce the need for set-up (Er, 2004). Set-up time, which occurs when different material options are processed in sequence in a certain process, is assumed to be constant for all processes and sequence independent.
- In the results reported here, each type of material has none, three or five different options, representing no variety, medium or a high level of variety.
- For any given level of product variety, all product variants have equal proportions over the total demand.

10.4.3 Simulation Factors

Three main factors are investigated in the simulation study. The first is *product variety*. The level of product variety produced by the manufacturer is determined by the number of options for main materials, unique material and packaging materials. The simulation experiments will investigate the impact of one, three and five material options to represent low, medium and high variety for main, unique and packaging material.

The second factor investigated is *supply lead-time uncertainty*, determined by the location and the reliability of suppliers. While supply lead-time uncertainty may also apply to companies operating in a purely national setting, the extent of supplier lead-time uncertainty used in the model captures the MNC context. In many MNC supply chains, transport of goods internationally is commonly done by sea, which is typically longer than local sourcing. Therefore, *local suppliers* are expected to deliver more quickly compared to *global suppliers* located in a different country. Ideally, the supplier delivers just in time. However, in reality, suppliers may deliver earlier or later than the scheduled time. The supply lead-time distribution is expected to be less variable if the supplier is local. With global suppliers, the supply lead-time distribution is affected not only by transportation times, but also potentially by additional delays due to problems associated with production, transportation schedules, customs and communication or unexpected problems due to weather, strikes, etc. (Levy, 1995). As a result, buying from global suppliers entails greater supply lead-time uncertainty, reflected in greater potential variability in supply lead-time.

The third factor to be investigated in this study is *demand uncertainty*, which is captured by the changes in the total volume of demand from one period to the next. Volume uncertainty is reflected in the deviation of demand from an average value. Demand uncertainty due to the continual updating of forecasts combined with long procurement time results in system nervousness, a problem that applies more commonly to MNCs than national companies (Houlihan, 1987). This is another characteristic of the MNC environment captured in the simulation.

In the model the actual demand in any period is determined by the forecast made in the previous period plus a random error that is normally distributed with mean zero. Then, the value of the forecast for each period is determined by the actual demand in that period multiplied by a constant value plus a minimum amount. The minimum amount is introduced to ensure that the forecasted demand will not have extreme values, e.g. a zero, or a negative demand level.

Results from the case companies indicate that forecasts predicting further into the future have higher uncertainty and higher possibility of errors. To incorporate this into the forecast generation procedure, the minimum amount (x) decreases as the forecast predicts further into the future.

The length of the forecast horizon depends on the supplier's lead-time. When suppliers are assumed to have a 3-month lead-time, then material for production in month 4 has to be bought in month one. This means that in month one, the forecast of demand in month 4 has to be available. If materials are bought from a supplier that can deliver in one month they only need to be bought based on the forecast made one month in advance.

10.4.4 Performance Measures

For this study we investigate the impact of product variety on the amount of time that a product spends in the system, i.e. *flow time*. Average flow time is a valuable metric to capture the responsiveness of an operational system (Kritchanchai and MacCarthy, 1999). Flow time is measured here by the period of time a product spends in the system from the time the demand for the product is initially generated by the internal customer until the product has completed manufacture. The flow time is measured in terms of its average values.

10.4.5 Design of Simulation Experiments

The general research question is divided into four specific questions to be investigated in the simulation study:

1. What is the impact on flow time of increasing product variety when both supply and demand are constant?

2. What is the impact on flow time of supply lead-time uncertainty in a high variety and constant demand situation?
3. What is the impact on flow time of demand uncertainty in a high variety and constant supply lead-time situation?
4. What is the impact on flow time of demand and supply uncertainty in a high variety situation?

First the simulation is run under *static conditions* with the system producing only one product variant and both supply lead-time and demand remaining constant. Then, the first set of experiments investigates the situation using different numbers of material options representing different levels of product variety whilst both supply lead-time and demand are constant. Comparing results from this set of experiments with *static conditions* will reveal the magnitude of impact of product variety on supply chain performance. Results from the first set of experiments are used as a *base case* to compare the results of subsequent sets of experiments. The next three sets of experiments investigate the situation with three and five material options under different demand and supply uncertainty situations. The second set of experiments investigates the situation when supply lead-time is uncertain but demand is constant. The third set of experiments investigates the situation when demand is uncertain and suppliers deliver materials with a constant lead-time. The final set of the experiments investigates the situation when both supply and demand are subject to uncertainty.

10.4.6 Input Data and Parameters

The simulation model strives to capture generic characteristics of MNC supply chains to give insights on the potential impact of different factors on their performance. It does not try to mimic any specific real system. Findings from the case companies are used to inform the simulation but specific quantitative data from companies has not been used, as it is not consistent with the objectives of the study. Instead, reference sets (Kritchanchai and MacCarthy, 2002) of input data and parameters were established that were representative of the generic observations in the empirical study. For example, three of the case companies provided some information regarding suppliers' lead-time. These companies generally did not know the exact form of supplier lead-time distribution, but they were able to estimate the minimum, maximum and most likely values. This indicated that sourcing from a local supplier on average required 20 days while global suppliers on average were expected to deliver in 60 days in such situations. A triangular distribution is commonly used to represent variability in such circumstances (Kelton *et al.*, 2003). A Normal distribution is not used, thus avoiding the possibility of meaningless zero or negative values.

The full reference sets of input data and parameters used in the simulation experiments are summarised in Tables 10.2 and 10.3. The input data that need to be determined are capacity, demand, processing time, set-up time and batch size. Production capacity in this model is based on typical working hours in most companies – 8 hours/day and 5 working days per week. Thus, the model has 20 working days or 160 working hours per month. The capacity available each month

Table 10.2. *The reference input data and parameters used in the simulation model*

Parameters	Values
Capacity	1 month = 20 days = 160 hours = 9600 minutes
Average demand	16875 units/month
Batch size	135 units of products
Processing time	0.0018 day/product = 0.864 minute/product
Process 1	0.0009 day/product = 0.432 minute/product
Process 2	0.00072 day/product = 0.3456 minute/product
Process 3	0.00018 day/product = 0.0864 minute/product
Set-up time	
Set-up process 1	0.75 day = 360 minutes
Set-up process 2	0.25 day = 120 minutes
Set-up process 3	0.0625 day = 30 minutes

is assumed to be constant. Demand that cannot be satisfied during a certain month is carried over as a backlog to the next month(s).

In setting the reference value of demand several points are considered:

- Findings from the empirical study indicate that, given the available capacity of 20 days per month, companies typically aim to produce a minimum of 15,000 units of product per month or 180,000 units of product per year.
- The model is designed to investigate the impact of having 1, 3 and 5 options for (1) main material, (2) unique material and (3) packaging material. Thus, to achieve a reasonable value (i.e. integer value) for product-level demand, the aggregate-level demand needs to be a multiple of 3 and 5.

These considerations suggest that the aggregate-level demand should be greater than 15,000 and the value should be divisible by 3 and 5. Taking into account the above considerations, the reference value for aggregate-level demand is set at 16,875 units per working month.

Based on the capacity and demand per month, the length of processing time can be determined. The system consists of three process stages - process 1, process 2 and process 3. The empirical results provide insights in determining the relative values of each stage of processing. As presented in Table 10.3, process 3 typically involves packaging that requires substantially shorter processing time compared to processes 1 and 2. Process 1 typically involves fabrication and manufacturing operations and thus requires relatively longer time than process 2. Thus, the lengths of process 1, 2 and 3 are set as 50%, 40%, and 10% of the total processing time, respectively.

Similarly, determination of setup time is informed by findings from the empirical study. The empirical evidence suggests that changes due to differences in main material are assumed to require major set-ups involving the entire production line, which typically requires up to 6–8 hours. Set-ups due to unique and packaging material changes will not require lengthy processes.

Another important parameter to consider is the batch size. The size of the batch plays an important role as it affects the amount of time the product spends in the system, i.e. flow time. The main criteria used to justify the size of batch are:

- Capacity utilization of process 1 (as the bottleneck process) has to be close to 75%.
- The batch size has to enable the entire product-level demand to be completed within the available capacity (20 days). In other words, in the deterministic situation the batch size has to prevent continuous backlogs, as it will lead to *system overflow*.

Based on these criteria, several trials have been conducted to determine an appropriate batch size. It was concluded that the smallest possible batch size (135 units), which meets all the considerations was the most reasonable value to adopt for the study. Further details on the determination of these data and parameters are discussed in Er (2004).

10.4.7 Tactical Planning for the Simulation Experiments

Simulation experiments require consideration of transient, dynamic and random effects. Before starting the experiments, it is important to ensure that the statistics

Table 10.3. *Factors and parameters used in the simulation model*

No	Factors	Level		
	Determinant of product variety	**Level 1 (Low variety)**	**Level 2 (Medium variety)**	**Level 3 (High variety)**
1	No. of different main materials	1	3	5
2	No. of different unique materials	1	3	5
3	No. of different packaging materials	1	3	5
	Supply Lead-time	**Level 1**	**Level 2**	
4	Main material supply lead-time	Local: Triangular (5, 20, 35) days	Global: Triangular (15, 60, 105) days	
5	Unique material supply lead-time	Local: Triangular (5, 20, 35) days	Global: Triangular (15, 60, 105) days	
6	Packaging material supply lead-time	Local: Triangular (5, 20, 35) days	Global: Triangular (15, 60, 105) days	
	Demand Uncertainty	**Level 1**	**Level 2**	
8	Demand	Constant = 16,875 units per month	Variable =' Demand = Forecast (i-1,1) + error N(0,3375) Forecast (i+1) = 13,500 + 0.2 x demand Forecast (i+2) = 10,125 + 0.4 x demand Forecast (i+3) = 6750 + 0.6 x demand Forecast (i+4) = 3375 + 0.8 x demand	

are collected when the simulation has reached a steady state (Law and Kelton, 2000). Visual examination of the results indicated that the system had reached steady-state conditions after approximately 5,000 warm up periods. Therefore, each experiment has been run for 5,000 warm up periods, followed by 5,000 measurement periods. In addition, each experiment conducted under uncertainty conditions has been replicated five times and average metrics estimated from the set of five replications.

10.5 Results from the Simulation Experiments

10.5.1 Static Condition

Results from the experiments under constant demand and supply conditions are shown in Table 10.4. In this set of experiments, the number of options for each material is increased from one to three and five, while the rest remain constant. Results from this set of experiments are used as base cases when comparing results obtained from subsequent experiments.

In the static situation when the system produces only one product and both supply and demand are constant, the average flow time is 7.78 days per product. The average flow time increases by 19% and 29% when the number of main material options is increased from one to three and to five, respectively. This is due to set-up activities that occur when different options of main material are processed consecutively in process 1. Increasing the number of unique and packaging materials to three and five, when there is one option of main material, does not result in a significant increase in the average flow time (less than 1%). These results are to be expected because of the assumptions made in the model. The set-up time associated with main material is significantly longer than the set-up time associated with unique and packaging materials.

The different impacts of main, unique and packaging material variety are also influenced by the fact that the effect occurs at different stages in production. Variety in main materials occurs at the beginning of the production process and the process itself takes a relatively long time (50% of total processing time), making main material availability critical to production. Thus, consequent delays associated with main material variety affect the entire production time of a batch and also subsequent batches. In contrast, packaging material variety takes place at the last stage of the production process. Thus, production of other batches in process 1 and process 2 are not blocked by extended lead-times due to packaging material variety.

Table 10.4. *Results from the static condition*

Measure	Unit of Measurement	1 Product	3 Options			5 Options		
			MM (Ex. 1)	UM (Ex.2)	PM (Ex.3)	MM (Ex. 1)	UM (Ex.2)	PM (Ex.3)
Average Flow Time	(Day/Product)	7.78	9.28	7.82	7.78	10.03	7.83	7.78

10.5.2 Supply Uncertainty Experiments

Results from the experiments for three and five material options with different levels of supply delivery uncertainty are summarised in Table 10.5. In this set of experiments, only delivery time for material(s) that is the source of product variety is changed from a fixed period of 20 days to a triangular distribution depending on the type of the supplier (local or global – see Table 10.3).

In analysing the results we will focus on three issues:

1. The impact of supply uncertainty.
2. The impact of different sources of variety and supply uncertainty.
3. The combined impact of increasing variety and supply uncertainty.

10.5.2.1 The Impact of Supply Uncertainty

Supply delivery uncertainty means that a certain amount of material may be delivered later than the expected time. Lateness in material delivery causes delays in production and eventually leads to longer average flow time.

As shown in Tables 10.6a and 10.6b, for the same level of product variety, increasing the level of supply uncertainty always leads to longer flow time. When the number of materials options is three, changing from suppliers that deliver in 20 days exactly (the base case) to ones that deliver within 5–35 days (local suppliers) results in increases of 4% to 17% in the average flow time depending on the source of product variety. A similar range is found when the number of material options is changed to five.

A much greater impact is evident when the level of supply uncertainty is increased. With both three and five material options, buying from global suppliers results in an increase of more than 30% in the average flow time compared to a constant supplier delivery time. This result suggests that increasing the level of delivery uncertainty results in much longer average flow times.

10.5.2.2 The Impact of Different Sources of Variety and Levels of Delivery

Tables 10.6a and 10.6b also show that different sources of product variety result in different impacts on the average flow time. As explained in the analysis of the static condition, main material variety has a higher impact on average flow time compared

Table 10.5. *Results from supply uncertainty experiments*

Measure	Unit of Measurement	Source of variety	3 material options		5 material options	
			Local Suppliers	Global Suppliers	Local Suppliers	Global Suppliers
Average Flow Time	(Days/Product)	MM	9.66	13.12	10.62	15.10
		UM	9.17	11.59	9.17	11.35
		PM	8.31	10.28	8.27	10.18

to unique or packaging material variety because the set-up times required to produce different types of main material are significantly longer.

Results from the experiment under supply uncertainty highlight that the timing of when materials are required for production is an important factor that contributes to the different levels of impact of different sources of variety on average flow time. In this experiment, suppliers for different types of material have the same supply delivery time distribution (5–35 days and 15–105 for local and global suppliers, respectively), which lead to a certain level of tardiness. However, for the same level of tardiness, different types of materials result in different magnitudes of delay in production. These findings are logical because the materials are required at different stages in the production. The main material is required at the beginning of the production, so lateness in main material delivery is likely to delay the entire production. On the other hand, packaging material is needed at the last stage of the production. This means that lateness in packaging material delivery does not necessarily cause delays in production because process 1 and process 2 can still begin. This result confirms the importance of main material availability for production and explains why main material variety has a much more significant impact on average flow time. The trends highlighted previously are amplified when supplier delivery uncertainty increases.

Although main material variety has the largest absolute effect on average flow time it should also be stated that in relative terms supply uncertainty results in greater increases for UM and PM compared to MM. Supply uncertainty amplifies the effect of other types of material, particularly UM, on average flow time. When UM are delivered on time, the average flow time is 7.82 day/product. When three options of UM are bought from local suppliers, the average flow time increase 17% to 9.17

Table 10.6a. *The impact of medium level of variety and different level of supply uncertainty on average flow time (day/unit product)*

Measure	Unit of Measurement	Source of variety	Base Case 3 Option	3 options			
				Local Suppliers	*Increase from base case (%)*	Global Suppliers	*Increase from base case (%)*
Average Flow Time	(Day/Product)	MM	9.28	9.66	4.06	13.12	41.33
		UM	7.82	9.17	17.20	11.59	48.13
		PM	7.79	8.31	6.74	10.28	32.04

Table 10.6b. *The impact of high level of variety and different level of supply uncertainty on average flow time (day/unit product)*

Measure	Unit of Measurement	Source of variety	Base Case 5 Option	5 options			
				Local Suppliers	*Increase from base case (%)*	Global Suppliers	*Increase from base case (%)*
Average Flow Time	(Day/Product)	MM	10.03	10.62	5.90	15.1	50.58
		UM	7.83	9.17	17.04	11.35	44.87
		PM	7.78	8.27	6.29	10.18	30.84

day/product. This means that uncertainty in UMs delivery time may cause a lot of disruptions and should also be managed carefully.

10.5.2.3 The Combined Impact of Product Variety and Supply Uncertainty

Figure 10.2 shows that there are interaction effects among different sources of product variety, the number of material options for each source of product variety and supply uncertainty. This can be seen in the different patterns for each of the lines shown. In general, changing the number of options for each source of variety from one to three to five, has less impact on the average flow time compared to changing the level of supply uncertainty. The magnitude of increase in average flow time due to increasing the number of material options for the same level of supply uncertainty is less than the magnitude of increase caused by increasing the level of supply uncertainty for the same number of material options. Increasing the level of supply uncertainty, especially when the number of options is high, results in significant increase in average flow time (30%–50%).

Different patterns are found for different combinations of material variety and supply uncertainty. For the same level of supply uncertainty, only increasing the number of main material options has a significant impact on the average flow time. Increasing the number of main material options from three to five leads to 10% and 15% increase in the average flow time when suppliers are local and global, respectively. Increasing the number of options of unique material or packaging material when the level of supply uncertainty is kept constant does not have a significant impact on the average flow time.

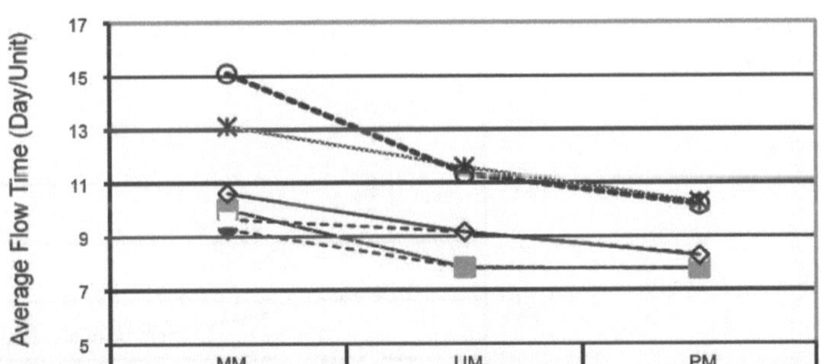

	MM	UM	PM
Base Case 3 Option	9.28	7.82	7.79
Base Case 5 Option	10.03	7.83	7.78
Local suppliers 3 option	9.66	9.17	8.31
Local Suppliers 5 options	10.62	9.17	8.27
Global suppliers 3 options	13.12	11.59	10.28
Global Suppliers 5 options	15.1	11.35	10.18

Source of Variety

Figure 10.2. *Average flow time when number of material options is three and five for different levels of supply delivery uncertainty*

Table 10.7. *Results from demand uncertainty experiments*

Measure	Unit of Measurement	Source of variety	Base Case 3 Option	Uncertain Demand with 3 options	*Increase from Base Case %*	Base Case 5 Options	Uncertain Demand with 5 Options	*Increase from Base Case %*
Average Flow Time	(Day/Product)	MM	9.28	15.91	*71.44*	10.03	24.58	*145.06*
		UM	7.82	11.87	*51.79*	7.83	12.47	*59.26*
		PM	7.79	11.19	*43.65*	7.78	11.36	*46.02*

10.5.3 Demand Uncertainty Experiments

Results from the experiments for a medium and high level of product variety with demand uncertainty are summarised in Table 10.7.

Demand uncertainty results from a degree of forecast error that eventually leads to a disparity between the material ordered and the actual production requirements that subsequently occur. On some occasions forecasts will overestimate actual demand. Thus, materials ordered by the manufacturer are greater than the actual production requirement. This means there are excess materials that have to be held for longer. Material shortages will occur when forecasts underestimate real demand. In the situation when current on-hand inventory is insufficient, the production of a batch will be delayed until sufficient materials arrive. In this set of experiments, it is assumed that there is no supply uncertainty, i.e. that suppliers deliver in fixed period of 20 days. This means production will be delayed for approximately 20 days until the next delivery arrives. This eventually leads to longer average flow times when there is demand uncertainty.

As shown in Table 10.7, demand uncertainty results in an increase of between 43% and 71% in average flow time when the number of material options is three. The magnitude of increase is greater (46%–145%) when the number of material options is five.

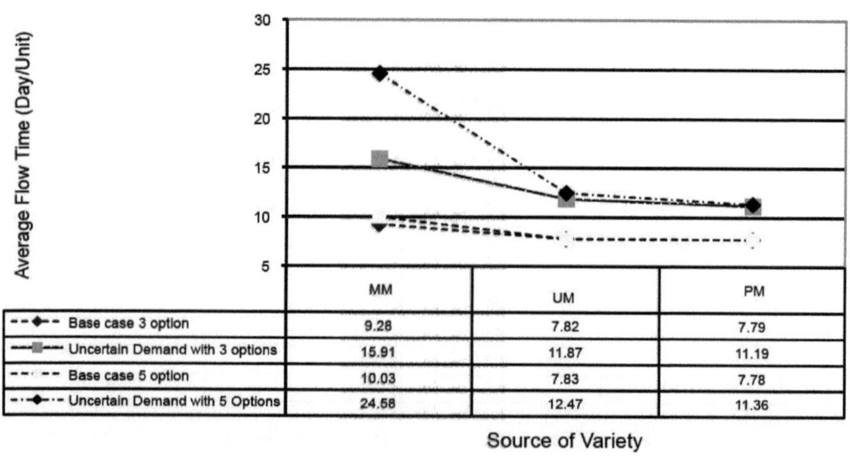

Figure 10.3. *Average flow time under demand uncertainty when number of material options is three and five*

Interaction effects are found between level and source of product variety and demand uncertainty. As shown in Figure 10.3, demand uncertainty clearly results in longer average flow time when the number of options is both three and five. Main material variety has the greatest impact on average flow time among different determinants of product variety in uncertain demand situations. This impact is more evident with the highest number of material options. Main material variety with demand uncertainty leads to increases in average flow time of the order of 145% when the number of options is five. However, the number of materials does not show a significant impact on the average flow time when the source of product variety is from unique or packaging materials. When the number of options is three and five, the magnitude of change in average flow time due to unique and packaging variety is 50% and 40%, respectively.

10.5.4 Supply and Demand Uncertainty Experiments

The previous experiments demonstrate that demand or supply uncertainty individually combined with product variety result in longer average flow times. In the presence of both sources of business uncertainty, the impact on average flow time is more pernicious. Disruptions caused by both supply and demand uncertainty clearly affect the average flow time. Supplier delivery uncertainty leads to occurrences of lateness in material delivery. Demand uncertainty means that when forecasts underestimate the actual demand, some materials are not available in the right amounts when required. Both of these situations cause delays in the production and eventually lead

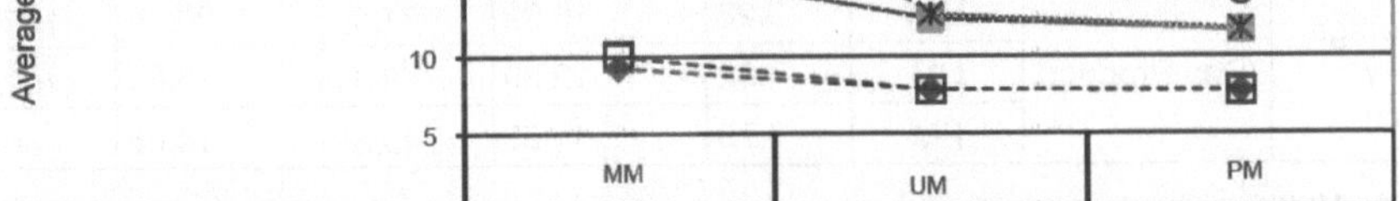

	MM	UM	PM
Base case 3 option	9.28	7.82	7.79
Demand uncertain, 3 opt & localsuppliers	15.92	12.39	11.65
Demand uncertain, 3 opt & global suppliers	20.7	15.81	14.05
Base case 5 option	10.03	7.83	7.78
Demand uncertain, 5 opt & local suppliers	26.86	12.66	11.74
Demand uncertain, 5 opt & global suppliers	32.9	15.66	14.09

Source of Variety

Figure 10.4. *Average flow time under supply and demand uncertainty when the number of material options is three and five*

to longer average flow times. Here we quantify the relative impacts of the combined sources of uncertainty.

As shown in Figure 10.4, every combination of supply uncertainty and demand uncertainty with a medium and a high number of material options results in longer average flow times relative to the base case.

10.5.4.1 The impact of supply and demand uncertainty on average flow time relative to base case

Tables 10.8a and 10.8b present the impact of different combinations of supply uncertainty and demand uncertainty in more detail with a medium and a high number of options on the average flow time.

As suggested previously, for every source of variety, demand and supply uncertainty have a pernicious impact on average flow time. As shown in Tables 10.8a and 10.8b, even when PM as the source of variety is bought from local suppliers, with the presence of demand uncertainty the average flow time increase is in the order of 50%. The condition is worst when UM and MM are the sources of variety where the increase in the average flow time is more than 60%.

When demand is uncertain, buying from global suppliers with greater supply uncertainty compared to local suppliers results in longer average flow times with both three and five material options. As shown in Table 10.8a, in the presence of demand uncertainty, buying materials from global suppliers results in an increase of 80% to 123% in the average flow time compared to the base case for a medium

Table 10.8a. *The impact of a medium level of variety, demand uncertainty and different levels of supply uncertainty on average flow time (day/unit product)*

Measure	Unit of Measurement	Source of variety	Base Case 3 Option	3 options			
				Demand Uncertain & Local Suppliers	*Increase from base case (%)*	Demand Uncertain & Global Suppliers	*Increase from base case (%)*
Average Flow Time	(Day/Product)	MM	9.28	15.92	*71.55*	20.70	*123.06*
		UM	7.82	12.39	*58.44*	15.81	*102.17*
		PM	7.79	11.65	*49.55*	14.05	*80.36*

Table 10.8b. *The impact of a medium level of variety, demand uncertainty and different levels of supply uncertainty on average flow time (day/unit product)*

Measure	Unit of Measurement	Source of variety	Base Case 5 Option	5 options			
				Demand Uncertain & Local Suppliers	*Increase from base case (%)*	Demand Uncertain & Global Suppliers	*Increase from base case (%)*
Average Flow Time	(Day/Product)	MM	10.03	26.86	*167.80*	32.90	*228.02*
		UM	7.83	12.66	*61.69*	15.66	*100.00*
		PM	7.78	11.74	*50.90*	14.09	*81.11*

number of options. For the same conditions, buying from local suppliers results in increases that range from 49% to 71%.

Similar trends are found when the number of options is changed to five. The only difference is that the impact of main material is much greater than when the number of options is three. For a high number of options, buying from global suppliers results in increases in average flow time that range from 81% to 228% under uncertain demand. Buying from local suppliers results in increases in average flow time that range from 50% to 167% for the same situation (Table 10.8b).

Finally, looking at the relative increase from base case highlights the interaction effects that exist among different sources of product variety, different number of material options, supply uncertainty and demand uncertainty. Changing the number of main material options in the presence of both uncertain demand and uncertain supply amplifies the different impacts caused by different sources of product variety. For every combination of supply and demand uncertainty investigated in this set of experiments, only increasing the number of options for the main material consistently results in longer average flow times.

Increasing the number of options for unique or packaging material does not have a significant impact on the average flow time. For a certain level of demand and supply uncertainty, the impacts caused by unique and packaging material variety on average flow time when the number of options is three and five are relatively similar. When material is bought from local suppliers, unique material variety results in 58% and 61% increases in the average flow time when the number of options is three and five, respectively.

10.5.5 Comparison of Overall Experiments

Having analysed results from each set of experiments, this sections compares the results from across all the experiments. Table 10.9a and 10.9b show the impact of different factors compared to the base case for medium and high level of variety, respectively. The table shows that in both medium and high level variety, demand uncertainty has a more significant impact compared to both the levels of supply uncertainty studied here.

For a medium level of variety, demand uncertainty increases the average flow time in the range 43% to 71% while the impact of supply uncertainty is within the range 4% to 6% (local suppliers) and 32% to 48% (global suppliers). When the number of options is five, demand uncertainty increases the average flow time in the range of 46% to 145%, while supply uncertainty is within the range of 6% to 17% for local suppliers and 30% to 50% for global suppliers.

These results are logical given the assumptions and parameters specified in the simulation model. Demand uncertainty influences the availability across the entire set of materials, while supply uncertainty affects only the availability of materials that are the source(s) of variety. This means that in the presence of demand uncertainty, the possibility of delays due to material lateness is higher, which eventually results in a longer average flow time.

Table 10.9a. *Overall findings in medium level of variety*

Measure	Unit of Measurement	Source of variety	Base Case 3 Option	Supply Uncertainty				Demand Uncertainty		Supply and Demand Uncertainty			
				Local Suppliers	*Increase from base case (%)*	Global Suppliers	*Increase from base case (%)*	Uncertain Demand with 3 options	*Increase from Base Case %*	Demand Uncertain & Local Suppliers	*Increase from base case (%)*	Demand Uncertain & Global Suppliers	*Increase from base case (%)*
Average Flow Time	(Day/Product)	MM	9.28	9.66	*4.06*	13.12	*41.33*	15.91	*71.44*	15.92	*71.55*	20.70	*123.06*
		UM	7.82	9.17	*17.20*	11.59	*48.13*	11.87	*51.79*	12.39	*58.44*	15.81	*102.17*
		PM	7.79	8.31	*6.74*	10.28	*32.04*	11.19	*43.65*	11.65	*49.55*	14.05	*80.36*

Table 10.9b. *Overall findings in high level of variety*

Measure	Unit of Measurement	Source of variety	Base Case 5 Option	Supply Uncertainty				Demand Uncertainty		Supply and Demand Uncertainty			
				Local Suppliers	*Increase from base case (%)*	Global Suppliers	*Increase from base case (%)*	Uncertain Demand with 5 Options	*Increase from Base Case %*	Demand Uncertain & Local Suppliers	*Increase from base case (%)*	Demand Uncertain & Global Suppliers	*Increase from base case (%)*
Average Flow Time	(Day/Product)	MM	10.03	10.62	*5.90*	15.1	*50.58*	24.58	*145.06*	26.86	*167.80*	32.90	*228.02*
		UM	7.83	9.17	*17.04*	11.35	*44.87*	12.47	*59.26*	12.66	*61.69*	15.66	*100.00*
		PM	7.78	8.27	*6.29*	10.18	*30.84*	11.36	*46.02*	11.74	*50.90*	14.09	*81.11*

Results from the third set of experiments appear to suggest that combined effects are not equal to the sum of the individual effects. For example, when the number of main material options is three, buying from local suppliers results in a 4.06% increase in the average flow time. When demand uncertainty is present the increase is 71.44%. When both factors are present the combined impact is 71.55%.

Nevertheless, the presence of all factors (variety, supply and demand uncertainty) clearly has a negative impact on the average flow time. This is particularly true for MM. When MM is the source of product variety, the presence of other factors amplifies the negative impact it has on average flow time. Figure 10.5 charts the impact of various conditions combined with main material variety on average flow time. For a medium level of variety, supply uncertainty increases the average flow time by 4% (local) and 41% (global). These increases rise to 71% and 120%, respectively, in the presence of demand uncertainty. The worst situation is when the number of main material options is five, materials are bought from global suppliers and demand is uncertain. In this situation the average flow time is more than trebled from 10.03 day/product in the base case to 32.90 day/product.

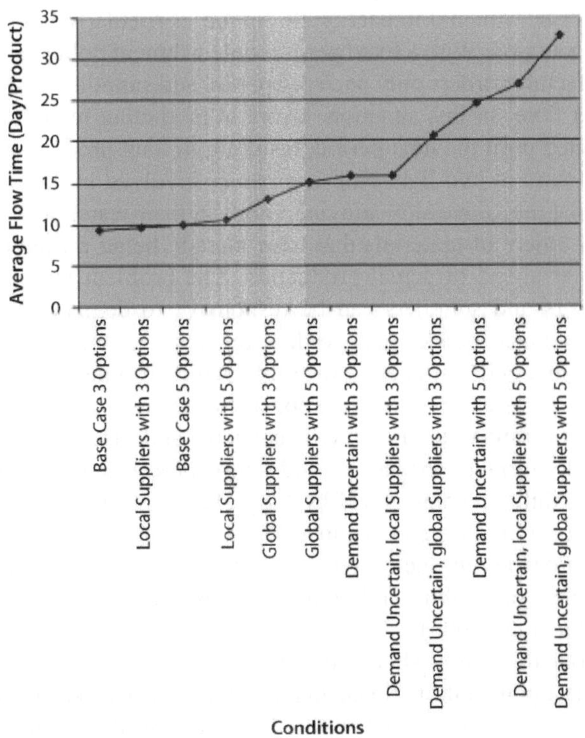

Figure 10.5. *The impact of main material variety and different factors on average flow time*

10.6 Practical Implications

Results from the simulation experiments have a number of practical implications. As expected, significant levels of product variety result in longer flow times. The simulation study shows that the detrimental impact of product variety is more pernicious for 'critical' materials, i.e. those required early in the production process and that entail long set-up times to change between different options. The simulation study has helped to identify the likely magnitude of these effects for some typical MNC supply chains. The simulation results also support the findings of Anderson (1995) and Thonneman and Bradley (2002) who have highlighted the detrimental impacts of setup on manufacturing performance.

The average flow time of a product is expected to be longer if critical materials are bought from suppliers with high lead-time uncertainty. The supply of such critical materials therefore requires a high service level to avoid unnecessary delays that lengthen flow times. Various strategies can be applied to achieve this, such as increasing safety stock and applying risk pooling techniques for critical materials (Lee, 2002), arranging service level agreements between manufacturers and suppliers, introducing penalty costs for late deliveries and strategically placing buffers after the most uncertain parts of the supply chain.

Companies facing demand uncertainty need to exercise care in high variety situations as it is more difficult to predict which materials to stock. The simulation results show that errors in predicting demand and which specific materials to procure may have a major detrimental impact on flow time, particularly when the materials are critical to production. With a fixed period replenishment policy as applied in this study, the manufacturer orders only once per period and suppliers will deliver within a fixed period of time. In this situation, errors in predicting material requirements cannot be corrected until the next period, resulting in long production delays. Such problems were found in two light bulb companies involved in the empirical study. They admitted that they have slow moving goods in their warehouse as they cannot adjust or amend orders of materials that have already being placed based on sales forecasts made far ahead of actual production. The problem may be mitigated if both manufacturers and suppliers can be responsive to demand changes, i.e. the manufacturer can place quick 'rush' orders to adapt to actual demand and the suppliers are flexible and fast enough to fulfil such rush orders. In an increasingly time sensitive business environment, responsiveness has emerged as an important competitive thrust for many industries (Kritchanchai and MacCarthy, 1999).

More generally, production planning and control processes in MNCs may develop rigidity and inflexibility, characterised by long planning lead-times and the use of forecasts generated a long time in advance of real demand. Successful introduction of fast track approaches in such environments requires careful consideration of the whole planning, scheduling and control process, its architecture and decision processes (Hamlin *et al.*, 2005).

Findings from this study should be of value in international operations and particularly in designing international manufacturing networks. Results from the simulation show that using a global supply base may result in significantly longer lead-times and higher inventory levels. Therefore, in dispersing their manufacturing activities companies should consider the availability of capable local suppliers, not merely basing the decision on cost or market drivers. This is highlighted by a shoe manufacturer in the empirical study that procures 70%–80% of materials from suppliers in Europe for production in Indonesia. Yet, this company has to cope with fluctuating demand and short product life cycles. Thus, they often have to expedite materials, late items or rush orders with air shipments, which are very costly. These facts highlight that without considering the availability of a capable supply base close to the manufacturing units, the cost advantages from cheap production might be negated by "hidden costs" such as express expedition, co-ordinating complex networks, customer cancellations due to late deliveries, etc.

Findings from the simulation study confirm the value of the concept of postponement, widely advocated in the literature (Lee *et al.*, 1993; Van Hoek, 1999). If variety proliferation can be postponed until later stages in production, disruption of production due to material tardiness can be minimized. Late product differentiation also allows more precise information on demand to be obtained, so mismatches between supply based on forecasts and actual production requirements can be reduced.

In designing or redesigning their international manufacturing networks, companies should also recognise that product variety does not necessarily mean introducing complexity across the entire network. Companies may reduce the negative impacts of product variety on supply network operations by choosing and applying strategies relevant to their product and process. Product-based strategies (Fisher *et al.*, 1999) that include the use of modular design, standardisation of materials and component sharing, may allow companies to offer high levels of product variety in the marketplace while maintaining a relatively low level of component variety and assembly complexity in production. Process-based strategies through the use of flexible technology and plant configuration based on the principles of cellular manufacturing may allow firms to accommodate a high level of variety at a reasonable cost. However, in order to successfully re-design products or processes in the supply chain, the impact on all functions needs to be considered and good co-ordination has to be maintained.

10.7 Concluding Remarks

Managing product variety that requires different types of materials in an international setting is very challenging. In this paper findings from a simulation study investigating the impact of product variety, supply lead-time and demand uncertainty on MNC supply chain performance have been presented. Results from the study show that increasing the number of materials options to three and five when such materials are critical for production results in increases in average flow time of the order of 20% and 30%, respectively, relative to static conditions. However, increasing the level of variety for materials that are required in later stages of production and requiring short set-ups does not have a significant impact on performance.

Offering any type of product variety in the presence of demand and supply lead-time uncertainty worsens the performance of a supply chain relative to a static situation. Sourcing a high variety of critical materials from suppliers with high delivery uncertainty results in increases in flow time of the order of 50%. Under uncertain demand scenarios, high levels of variety in critical materials result in increases in flow time of the order of 140%. The worst performance is found when the system has to handle critical material variety under both demand and supply uncertainty. Increases in flow time of the order of 200% may occur. These results highlight the need for careful management of variety in international operations.

The findings from the study provide important insights on the magnitude of the impact of product variety in the context of international operations. Opportunities exist for further studies. This study has concentrated on three levels of variety only. Adding more levels of product variety could provide more insights on the impact of increasing product variety on performance of a supply chain. Further work might also investigate the 'shape' of the lead-time distributions in addition to the level of variability. In this study, the international dimension has concentrated on upstream planning and production activities. Incorporating downstream distribution, logistics and transport activities may add to our understanding of international supply chain management. Supply networks with multiple manufacturing sites and multiple potential supply routes are also of increasing importance. Although simulation in this wider context is challenging it may provide an understanding on how flexibility in international production networks can be exploited to produce higher levels of variety.

References

Akkermans, H., Bogerd, P. and Vos, B. (1999), Virtuous and Vicious Cycles on the Road Towards International Supply Chain Management, International Journal of Operations & Production Management, Vol. 19, No. 5/6, pp. 565–581.

Al-Zubaidi, H. and Tyler, D. (2004), A Simulation Model of Quick Response Replenishment of Seasonal Clothing, International Journal of Retail & Distribution Management, Vol. 32, No. 6, pp. 320–327.

Anderson, S.W. (1995), Measuring the Impact of Product Mix Heterogeneity on Manufacturing Overhead Cost, The Accounting Review, Vol. 70, No. 3, pp. 363–387.

Bils, M., and Klenow, P.J., (2001), The Acceleration in Product Variety, The American Economic Review, Vol. 91, No. 2, pp. 274–280.

Cox, W.M., and Alm, R., (1998), The Right Stuff: America's Move to Mass Customization, Federal Reserve Bank of Dallas.

Er, M. (2004), Managing Product Variety in International Supply Chains, Ph.D Thesis, University of Nottingham, UK.

Er, M. and MacCarthy, B.L. (2002), Configuration of International Supply Networks and their Operational Implications: Evidences from Manufacturing Companies in Indonesia, Proceedings of the 7th Cambridge International Manufacturing Symposium, Cambridge, 12–13 September 2002, pp. 60–71.

Ferdows, K. (1989), "Mapping International Factory Network", in Ferdows, K. (ed.), Managing International Manufacturing, North-Holland, New York, pp. 3–21.

Fisher, M.L., Hammond, J.H., Obermeyer, W.R., and Raman, A. (1994), Making Supply Meet Demand in an Uncertain World, Harvard Business Review, Vol. 72, No. 3, pp. 83-93.

Fisher, M.L., and Ittner, C.D. (1999), The Impact of Product Variety on Automobile Assembly Operations: Empirical Evidence and Simulation Analysis, Management Science, Vol. 45, No. 6, pp. 771–786.

Fisher, M.L., Ramdas, K. and Ulrich, K. (1999), Component Sharing in the Management of Product Variety: A Study of Automotive Braking Systems, Management Science, Vol. 45, No. 3, pp. 297–315.

Ghoshal, S. and Bartlett, C.A. (1990), The Multinational Corporation as an Interorganizational Network, Academy of Management Review, Vol. 15, No. 4, pp. 603–625.

Hamlin, M., MacCarthy, B.L. and Guinery, J., (2005), Fast Track Order Fulfilment at Henkel: the PROCHART approach, Institute of Operations Management CONTROL Magazine, Vol. 31, No. 4, pp. 15–19.

Houlihan, J.B. (1987), International Supply Chain Management, International Journal of Physical Distribution and Materials Management, Vol. 17, No. 2, pp. 51–66.

Kelton, D.W., Sadowski, R.P. and Sturrock, D.T. (2003), Simulation with Arena, Third Edition, McGraw Hill, Singapore.

Kritchanchai, D. and MacCarthy, B.L. (2002), A Procedure for Establishing a Reference State in Qualitative Simulation of Operational Systems, Industrial Management & Data Systems, Vol. 102, No. 6, pp. 332–340.

Kritchanchai, D. and MacCarthy, B.L. (999), Responsiveness of the Order Fulfilment Process, International Journal of Operations & Production Management, Vol. 19, No. 8, pp. 812–833.

Law, A.M. and Kelton, W.D. (2000), Simulation Modelling and Analysis, Third Edition, McGraw-Hill, Singapore.

Lee, H. L. (2002), Aligning Supply Chain Strategies with Product Uncertainties, California Management Review, Vol. 44, No. 3, pp. 105–119.

Lee, H.L., Billington, C. and Carter, B. (1993), Hewlett-Packard Gains Control of Inventory and Service through Design for Localization, Interfaces, Vol. 23, July–August, pp. 1–11.

Levy, D.L. (1995), International Sourcing and Supply Chain Stability, Journal of International Business Studies, Second Quarter, pp. 343–60.

Lowson, R.H. (2001), Offshore Sourcing: an Optimal Operational Strategy?, Business Horizon, November–December, pp. 61–66.

MacCarthy, B.L. and Atthirawong, W. (2003), Factors Affecting Location Decisions in International Operations: a Delphi Study, International Journal of Operations & Production Management, Vol. 23, No. 7, pp. 794–818.

MacDuffie, J.P., Sethuraman, K., and Fisher, M.L., (1996), Product Variety and Manufacturing Performance: Evidence from the International Automotive Assembly Plant Study, Management Science, Vol. 42, No. 3, pp. 350–369.

McGrath, M.E. and Bequillard, R.B. (1989), "International Manufacturing Strategies", in Ferdows, K. (ed.), Managing International Manufacturing, North-Holland, New York, pp. 23–40.

Milgate, M., (2001), Supply Chain Complexity and Delivery Performance: An International Exploratory Study, Supply Chain Management: An International Journal, Vol. 6, No. 3, pp. 106–118.

Minuteman Software (2005), GPPS World™, http://www.minutemansoftware.com/features.htm

Oliff, M.D., Arpan, J.S. and DuBois, F.L. (1989), "Global Manufacturing Rationalisation: The Design and Management of International Factory Networks", in Ferdows, K. (ed.), Managing International Manufacturing, pp. 41 – 65, North-Holland, New York, NY.

Ovalle, O.R. and Marques, A.C. (2003), The Effectiveness of Using e-collaboration Tools in the Supply Chain: an Assessment Study with System Dynamics, Journal of Purchasing and Supply Management, Vol. 9, No. 4, pp. 151–163.

Persson, F. and Olhager, J. (2002), Performance Simulation of Supply Chain Designs, International Journal of Production Economics, Vol. 77, No. 3, pp. 231–245.

Porter, M.E. (1986), Changing Patterns of International Competition, California Management Review, Vol. XXVIII, No. 2, pp. 9–40.

Ramdas, K., (2002), Managing Product Variety: An Integrative Review and Research Directions, Production and Operations Management, Vol. 12, No. 1, pp 79–101.

Randall, T. and Ulrich, K. (2001), Product Variety, Supply Chain Structure, and Firm Performance: Analysis of the U.S. Bicycle Industry, Management Science, Vol. 47, No. 12, pp. 1588–1604.

Spengler, T. and Schroter, M. (2003), Strategic Management of Spare Parts in Closed-Loop Supply Chains – A System Dynamics Approach, Interfaces, Vol. 33, No. 6, pp. 7–17.

Thonemann, U.W. and Bradley, J.R. (2002), The Effect of Product Variety on Supply-Chain Performance, European Journal of Operational Research, Vol. 143, No. 3, pp. 548–569.

Tiger, A.A. and Simpson, P. (2003), Using Discrete-Event Simulation to Create Flexibility in APAC Supply Chain Management, Global Journal of Flexible Systems Management, Vol. 4, No. 4, pp. 15–22.

Towill, D.R. (1996), Industrial Dynamics Modelling of Supply Chains, International Journal of Physical Distribution and Logistics Management, Vol. 26, No. 2, pp. 23–42.

Van Hoek, R.I. (1999), Postponement and the Reconfiguration Challenge for Food Supply Chains, Supply Chain Management, Vol. 4, No. 1, pp. 18–34.

Wisner, J.D., Leong, G.K. and Tan, K.C. (2004), Principles of Supply Chain Management: A Balanced Approach, Thomson South-Western, Mason, OH.

Wyland, B., Buxton, K. and Fuqua, B. (2000), Simulating the Supply Chain, IEE Solutions, January, pp. 37–42.

11 Set-Up and Operation of Global Engineering Networks Spanning Industrialized and Emerging Economies

Harshavardhan Karandikar
ABB Inc. (Power Products Medium Voltage North America)

Abstract The practice of technology and product development in most industries today relies upon global engineering networks (GEN) where members of the multi-functional development team come from many locations and organizations. More recently, there has been a clear shift to using development resources from the so called "Emerging Countries" (ECs) that have fast growing domestic markets, large pools of skilled workforce and low labor costs. The integration of resources from ECs, whether from the company's own businesses there or from external suppliers, poses a new challenge for collaborative work in a GEN. These challenges are described in this chapter. The process of transition to the use of resources from ECs is fraught with business risks. These risks are described together with steps than can be taken to mitigate them. A model, derived based on practice, to manage the transition process is presented. The model is comprised of a set of Foundation elements and location-dependent Transformation processes. This transition model has continued application during the post-transition operational phase of the GEN. The observations and conclusions presented here are supported by four cases studied over the past five years.

Keywords Distributed development, Global engineering, Multi-functional development team, Virtual teams

11.1 Introduction

It is common practice in many industries today to carry out product development, from concept to first prototype, using geographically distributed teams. This is also done for technology development and R&D activities. This is particularly the case with large corporations that develop complex products or large systems where all the necessary competencies may not be available at one location. A 2006 survey by the Offshoring Research Network, see Couto *et al.* (2006), claims that the practice of global engineering and product design is growing at a rate greater than 50%. Further details of the global engineering trends are provided in Dehoff and Sehgal (2007) and a broad examination of the global product development practice is presented in Eppinger and Chitkara (2006).

There is extensive academic literature covering various aspects of global R&D and product development. A number of definitions of the so called virtual teams are listed in Powell *et al.* (2004). Snow *et al.* (1992) describe an organizational form called "network organization" that is a cluster of business units coordinated by market mechanisms. The term Global Engineering Network (GEN) is used to describe an organizational form that falls between that of a virtual team and a network organization. It corresponds closely to the Multinational Work Teams of Earley and Gibson (2002). In contrast to a virtual team, the relationship between the partners in the case of a GEN is longer term and may span the operation of many virtual teams. Furthermore, the team membership is from different organizations where each organization has separate business targets but brings a specific competitive advantage to the table. Thus, technical as well as business, commercial and legal aspects must be actively and simultaneously considered when setting up the collaboration. In contrast to a network organization, middle-management planners have a critical role to play in GEN; the set-up of the network is not left purely to market forces. An integrating framework, comprising context, capabilities and configuration, and key patterns of GEN are covered in Zhang *et al.* (2007).

In this chapter, we focus our attention on a particular type of GEN, one where the membership is a combination of resources from countries with advanced industrial economies, e.g., the western countries, and those from emerging economies such as China, India and countries of Eastern Europe. The former are referred to as industrialized countries (ICs) and the latter as emerging countries (ECs). The ECs offer three main advantages:

1. Fast growing domestic market: Local and regional demands create opportunities for setting up manufacturing in ECs. As a result, migrating product development and engineering to ECs in order to incorporate local know-how also becomes an attractive prospect.
2. Availability of workforce: Depending on the domain and discipline the availability of sufficient numbers of highly skilled and competent technical resources is a big advantage. This is especially the case for mature technical disciplines where the output from the universities has been declining in the ICs.
3. Low labor cost: Highly competitive wages and low cost compared to ICs make engineering in ECs profitable and a competitive necessity. However, cost can be a temporary advantage and cannot be the sole basis of a long-term strategy for distributing the engineering effort.

Thus, it is only natural for global engineering and manufacturing firms to utilize opportunities created in ECs as is being done by many of them as part of their "cost migration" strategies (Vestring *et al.*, 2005). An additional advantage of such a global network is that of flexibility (Stock *et al.*, 1998), changeability (Dekkers and van Luttervelt, 2006) or agility (Lee and Lau, 1999). This is needed to quickly respond to windows of market opportunity around the world. In fact, Parker (1999) cites speed as one of the prime motivations for collaborative development efforts with suppliers. In addition to increased speed to market, Couto *et al.* (2006) cite access to qualified personnel as an increasingly important key reason to move to global offshoring.

11.2 Review of Literature on Distributed Development

The challenges faced by virtual teams have been extensively researched and an excellent review is provided in Powell *et al.* (2004). They also point out the lack of published research on global, long term virtual teams – the kind that concern us here. The extent of use of global virtual teams is reported by McDonough *et al.* (2001). The principal rationale for the use of such teams is stated to be the need to understand the customers in new, emerging and fast growing markets. The desire to leverage scattered expertise is the next objective stated by the companies employing such teams. The presence of behavioral and managerial challenges for such teams is also identified. When the team is distributed over time zones with large chunks of non-overlapping working hours the complexity of management and coordination increases even more, as discussed by Mar-Yohanna (2001). Specifically, he discusses issues of leading global project teams. Vinaja (2003) stresses the multi-cultural issues in virtual teams and cultural factors that impact leading such teams and the influence on the nature and strength of collaboration. Additional challenges arise when the team is composed of individuals from different organizations – within the same company or from different companies, e.g., suppliers. The difficulties of melding together teams comprising members from different profit centers within the same company, as was the case with the teams studied in Karandikar and Nidamarthi (2006), should not be underestimated. An extensive study of so-called multinational work teams, including a theoretical model incorporating their key features, is presented in Earley and Gibson (2002). The integration of suppliers in new product development or engineering has also been deeply studied and there exist best practices on how to go about doing this.

Two main sets of aspects, relationship structuring and asset sharing, are pointed out in Ragatz *et al.* (1997) as critical to supplier integration in new product development. Bevan (1989) proposed the concept of "co-makership", implying a strongly collaborative and strategic relationship with suppliers, and showed how it enhanced competitiveness. Different modes of global product development, based on ownership and location of resources, are described in Eppinger and Chitkara (2006). These are Captive Offshore, Global Outsource, Local Outsource and Centralized. However, one element that is not explicitly mentioned but is critical is the nature of the contractual relationship between the various organizational components of the GEN. Thus, one of the cases presented here later covers a mode that is Captive Offshore but uses a contractual relationship of Global Outsource. This is necessitated by one of the challenges facing GEN and mentioned in the next section, i.e., the difficulty in sharing risks and rewards among different profit centers within a large global company when the components of GEN come from these different profit centers.

The problems of geographic distribution and how they can be overcome using information technology is reported with examples in Boutelier *et al.* (1997), Reddy *et al.* (1993), and Nowak *et al.* (2002). Most virtual team literature, particularly related to product development, has dwelt on team sizes of less than 10 persons. The teams described here were always larger and typically of the order of 15-40 persons spread across 3–5 locations, countries, and organizations. The organization of work in such teams poses a higher level of management challenge. However, such situations are not uncommon in large product development projects that involve suppliers. This has certainly been the case in aerospace and automotive industries. In these industries it has long been the practice to partition the overall development effort and to parcel out well defined development tasks with clear deadlines. Thus there exists a body of knowledge in practice of how to set up such teams and how to select the organizational partners, i.e., suppliers, to participate in the development (Ragatz *et al.*, 1997). Similar experience exists in software development driven by the off-shoring trend of the last decade. However, we believe that the challenge of executing engineering activities in a network is much greater than in the case of software due to the difficulty of transferring and sharing implicit knowledge and the importance of domain know-how.

In addition to the complexity factors mentioned earlier other significant ones have arisen in recent times, namely, when work is spread across locations with widely differing wage spectrums and the approach to intellectual property (IP) protection. This, in particular, is an issue today as many companies are trying to reduce their cost of development by conducting some of it in ECs. This does not appear to have been addressed in existing literature though with some exceptions. Grabowski *et al.* (2001) discuss experience from joint European-Chinese developments where the principal need was to locate, understand and protect the know-how existing at different locations. The latter should not be an issue when the GEN primarily comprises members from one global organization, though it is in the case of IC-EC networks due to the insecurity faced by certain members of the network concerning their jobs. What has also not received attention are the challenges encountered in the process of transition from localized to distributed development particularly when the new locations added are from ECs. These issues are covered in Karandikar and Nidamarthi (2006) and are briefly presented here once more.

11.3 Challenges Facing GENs

The operational challenges facing GENs have been reported in the context of virtual teams in the literature cited earlier. Jonsson *et al.* (2001) categorize these challenges as geographical (time and distance separation hindering communication), cultural (work habits, differences in language proficiency leading to misunderstanding, etc.) and organizational. In Karandikar and Nidamarthi (2006) the challenges are presented as shown in Figure 11.1. The problems of distance, time, structure, and culture have been extensively covered in the virtual team literature. However, in case of GENs comprising IC and EC entities, the problems thrown up by economic factors and strategy are special.

One of the issues faced in a GEN as well as in sourcing networks in ECs is the concern about protection of intellectual property (IP). When dealing with advanced engineering and design activities there is a concern of the loss of this know-how outside the core firm. There exist clear guidelines and practices on how to deal with this when dealing with suppliers. However, no such guidelines exist for internal engineering networks even though the dangers may be just as large. While there is

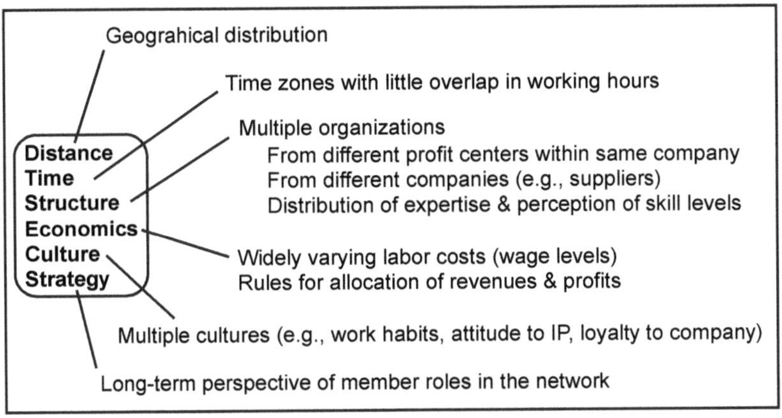

Figure 11.1. *Challenges facing teams in a GEN*

little grounding in fact there is a strong perception among engineers in the ICs that their counterparts in ECs will easily defect, along with the know-how, to competition in the EC markets. As discussed by Dietz *et al.* (2005), it is possible to take clear strategic and operational measures to overcome these problems rather than relying on legal measures with a questionable record of success.

A second issue, peculiar to GENs comprising members from ICs and ECs, is the perceived threat to the IC location member of the loss of work to the EC locations in the medium to long term. This is particularly critical and a major business risk in the start-up phase of such a GEN. The members in ICs perceiving poor long-term career prospects may start leaving the organization thus resulting in a crucial loss of competency. Such risks have been reported by Bloch and Jans (2005) in the context of software off-shoring where a high rate of turnover at the onshore location threatened the transfer of knowledge to the EC location. This can doubly hurt because the transfer of high-end technical work, where competence is developed over many years, is a gradual process. The mitigation of this risk through clear communication and a sound and transparent strategy for long-term work allocation is discussed later.

11.4 A Model for the Set-Up of GENs

The previous section highlighted the problems faced in the operation of GENs and also described a significant business risk during the launch phase and the transition phase of a GEN. Once a strategic decision has been made to move from a localized development effort to GEN – a global, virtual, multi-cultural, multi-organizational team – the transition process has to be managed in a balanced way. A model, shown in Figure 11.2, has been proposed to guide this process. The model comprises three foundational elements and three transformational processes.

This model arose out of the author's deep involvement and participation in at least four instances when such GENs were established. Three of these cases were presented first in Karandikar and Nidamarthi (2006). The approximately five-year

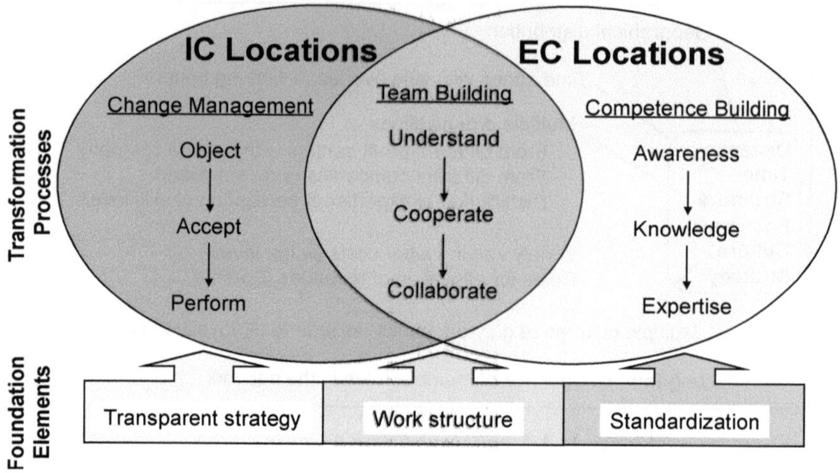

Figure 11.2. *A model for the management of the GEN transition process*

duration covering these cases, starting in Spring 2003 and with Case 4 initiated in the summer of 2007, a study of the literature cited here and post-mortem analysis of the progress made in each case has increased the degree of confidence in the relevance and applicability of this model.

Existing teams in ICs undergo a transition from insecurity, often manifesting itself as objection and passive resistance to EC team participation, to accepting EC colleagues, and finally reaching peak performance in the collaborative environment. The strategy here is to follow basic principles of change management – clear communication of objectives and rationale, address the insecurity and fears, and motivation and coaching combined with championing. The teams in ECs need to be developed from novice to expertise level (in terms of domain specific engineering work). This competence build-up needs to be achieved from simplified engineering to complex jobs. Standard engineering solutions and processes help manage this transition. The management challenge in this transition phase is to bridge the two sets of teams, in IC and EC, and meld them towards common goals. During the transition phase, both sets of teams transform from understanding to collaborating as if familiar colleagues at one location are working together. Here, a clear work structure and robust project management have to be employed in conjunction with the transformation processes at the IC and EC locations.

The GEN transition model shares some commonalities with the model for supplier integration presented in Petersen *et al.* (2003), if we treat the various development locations as individual suppliers. The transition model takes into account the following aspects of the supplier integration model, in particular, via the team building transformation element:

- Increased knowledge of the supplier is more likely to result in greater information sharing and involvement of the supplier in the product development process.
- Sharing of technical information increases levels of supplier involvement and improves outcomes.
- Supplier involvement leads to higher achievement of team goals.

- Problems associated with technical uncertainty can be mitigated by greater information sharing and supplier participation.

The transition model complements the conceptual framework for multinational teams described in Earley and Gibson (2002). They describe individual and group elements linked through specific processes that are inherent in teamwork. Further, they describe certain catalyst elements, namely, social structure and work structure, which provide the impetus for the processes and support them.

11.4.1 Foundation Elements

The most critical aspect for ensuring a successful transition and the long-term stable operation of a GEN is to make the strategy for global development and the rationale for it transparent and to explicitly communicate it. The reasons for the GEN strategy are typically fourfold:

1. The need to set up a development presence in the EC in order to better understand the market and to serve it on cost competitive terms.
2. A lack of sufficient numbers of competent resources at a single location.
3. The need to reduce overall development costs without sacrificing the number and quality of product development targets.
4. A gain in flexibility of allocation of resources. This is especially important in a project oriented or an engineer-to-order product business where there can be large variation in the engineering workload.

A second, equally critical aspect of transparency and one related to the second element, work structure, is to communicate the respective long-term roles of the members of the GEN and the consequences of the role distribution. This implies that there is a clearly foreseen distribution of work between the IC and EC locations. A lack of such organization of effort or the failure to communicate the planned allocation can have the disastrous consequence of a rapid loss of competent personnel at the IC locations. The communication is needed to counter unfounded fears and to rationally deal with well-founded concerns.

Considering the difficulty of carrying out complex engineering and development tasks in a project-based or engineer-to-order product business in a network, critical attention also needs to be paid to better definition of the engineering content through standardization. The benefits of doing this have been described in Hoare and Seiler (2001) and Eppinger and Chitkara (2006), who identify this as product and process modularity. As mentioned in the latter, much more than component and part standardization is needed to ensure effective collaboration. The scope of standardization should cover standard solution concepts and design rules (see Cases 1 and 3 later) and the development of common and shared work processes including common engineering analysis, design calculation, and product data management tools. Such standardization reduces the chances of miscommunication and also unnecessary design effort thus improving the efficiency of the engineering process. Engineering can be both optimized and globalized by means of standard solutions that can be repeatedly used or easily scaled in customer projects. That is, a system is delivered by using as many standard solutions that are common across countries.

All work processes within the business – sales, engineering, and supply – need to be re-defined based on standards and should be globally optimized. For example, cost effective suppliers from ECs can be developed for supplying complete standards to all global engineering locations.

11.4.2 Transformation Processes

The three transformation processes have to run in parallel during the transition process. The processes are tailored to the location in the network, whether it is an EC or IC location. The first of these is a change management at the IC sites focused on breaking down any resistance to the spread and transfer of development activities to EC locations. Organizational culture needs to change from local, independent, and custom engineering to global, shared, and standards based engineering. The second process, which is run across all locations, is to build up a strong team and is focused on improving the level of trust and increasing the collaboration among the various sites. Finally, the third process is focused on building up competence and expertise at the new EC development sites.

The typical reaction at the IC locations of the decision to set up the GEN with participation of EC locations is at first to vehemently object to the move on grounds of the unique nature of the technical know-how and the depth of expertise that is needed to master the engineering tasks. This is particularly the case with engineer-to-order products and systems. Overcoming this resistance and laying the groundwork for long-term and frictionless collaboration is the aim of the change management transformation process. One critical requirement here is to have individuals who will champion the GEN within the IC locations. Furthermore, based on our experience, the GEN must be seen as an opportunity for innovation and cost-competitive entry into newer markets rather than as a threat to jobs at IC locations.

Two of the critical factors for successful operation of GEN are mutual trust, among network participants, in their ability to execute the allocated tasks and the corresponding appreciation of the competence of the members at the other locations. The critical influence of credibility and trust on the performance of a distributed team is also dealt with in Benson-Armer and Hsieh (1997). It is important to note a point made in Ragatz *et al.* (1997) that trust is developed more through performance to expectations over time rather than any trust building or team building interventions.

Thus, time has to be allowed in the transition phase for the training of resources and the build up of competence at EC locations. This also requires the development of a clear training plan that may involve rotation of some members from the EC locations to the IC locations for a period of weeks or months. In the case of working with suppliers the required expertise typically already exists as this is normally a precondition for the selection of that supplier. The competence build-up is facilitated by a number of measures such as,

- allocation of simpler tasks that are then gradually increased in complexity;
- rotation of key technical personnel between the different locations; and
- availability of and ready access to documented engineering know-how, including standard designs.

In the case of engineering, the competence in question includes adequate domain know-how as well as sufficient understanding of the typical customer problems and the ability to convert customer requirements or internal specifications into workable and manufacturable designs.

Within Team Building, the transition from cooperation to collaboration requires a more active participation by all entities involved and a significant commitment of time and resources. The exchange of information and ideas becomes bi-directional rather than the EC locations looking for management as well as technical direction from the IC locations.

11.5 A Note on GEN Operation

There is a considerable body of literature on the management and operation of GENs and thus this aspect is not covered here in detail. As an example, Eppinger and Chitkara (2006) describe a number of operational practices, success factors, as well as best practices. Many of the sources cited in the literature review earlier cover the issues in the context of virtual or distributed teams.

The foundation elements of the GEN transition model continue to be important in the operational phase of the GEN. Furthermore, due to the dynamic nature of the environment in which GENs operate and in order to ensure smooth functioning a distributed team must always "close the loop". As shown in Figure 11.3, there is the continuous larger loop of planning, communication, and coordination activities. However, even within each activity it is important to ensure that the message delivered from one location has been understood at the other locations, i.e., the intent of the communication is clear to everyone in the network to whom it is directed to. Thus there are these smaller loops. It often happens that communication within teams gets done as a "broadcast", usually using e-mail. For the smooth operation of a multi-functional, multi-cultural, distributed team, as is the case with GENs, one must ensure that this message has been received and fully understood at the other, especially, remote locations. This is called "Closing the Loop". If not done it almost always leads to future misunderstandings and contributes to project or development

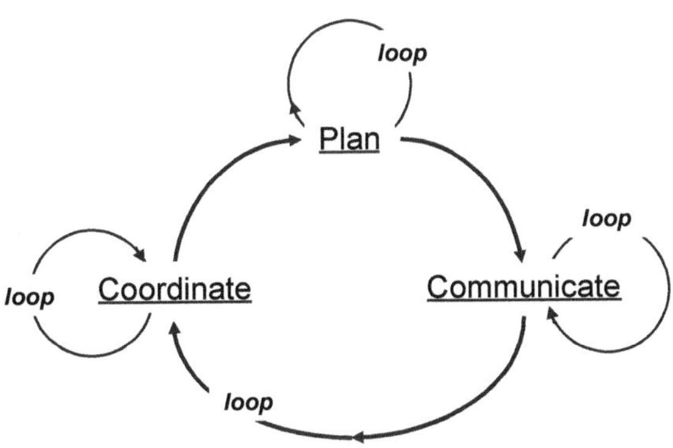

Figure 11.3. *GEN operation: closing the loops*

snafus. This principle must be applied to both formal and informal communications within GENs.

11.6 GEN Cases

The model proposed earlier is supported by four cases where the author was involved in the transition to GEN over the past five years. The cases are described next, their scope is shown in Figure 11.4 and their characteristics are summarized in Table 11.1.

11.6.1 Case 1 – Manufacturing Automation Systems Business

The Manufacturing Automation Systems business studied is engaged in designing, producing, installing, and commissioning, assembly lines for the manufacture of automotive powertrain components, e.g., gearboxes, axles and engines. The assembly line is typically comprised of a transport mechanism, e.g., a conveyor, and a series of stations or robotic cells such that one or more manufacturing operations are executed at each station or cell. This is a project driven business where the number of projects that may be executed per year is in the tens. The business itself is operated from four locations, each with a different system and market focus though with some overlaps. Three of the four locations have a long history. They have been close to the customer base and have accumulated significant technical competence over the course of time. However, they are all in ICs with high total wage burdens. The fourth location, in an EC, has been recently set up due to significant growth potential in this market, thus necessitating the need to be close to the new customer base. This also has the added benefit of a lower cost base for the operations. The nature of the business is such that engineering and manufacturing assembly and installation labor hours are significant components of the total project cost, and thus the lower wage burdens of the EC locations can be an advantage provided the same or better level of efficiency as the IC location can be achieved.

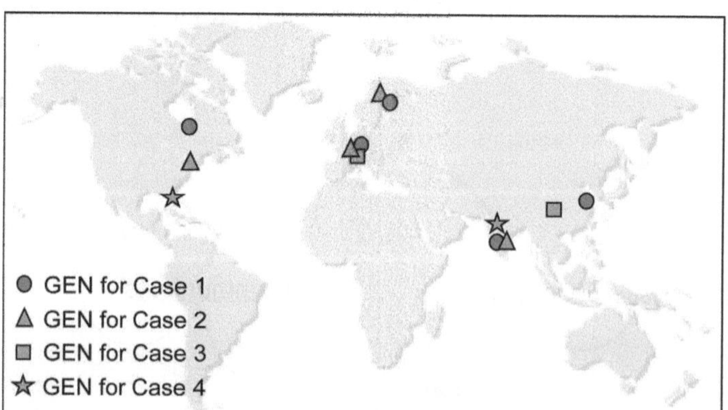

Figure 11.4. *Global Engineering Network for the three cases*

A strategic decision was made by the senior management to increase the amount of engineering and component sourcing to be done from the EC locations. With growing business in EC regions this made strong business sense. The supply management function was actively involved in identifying and qualifying the supplier base and in setting up strategic partnerships.

In the case of engineering, an analysis was carried out to identify non-core basic design activities that were then to be carried out at the EC location. It was not possible to do all engineering at the EC location in the short term due to the lack of relevant technical design competence that took two years to build up. Additionally, an external supplier was also used at another EC location to carry out basic mechanical design and detailing tasks. A strategic agreement was signed with this supplier outlining the kinds of activities they will be responsible for in the future and the expected

Table 11.1. *Characteristics of the cases*

	Case 1	Case 2	Case 3	Case 4
Characteristic of the GEN (all with both IC & EC locations)	• Global with largely non-overlapping work hours • Multi-cultural (5 countries) • Multi-company, Multi-organizational – 4 units from same company and an external supplier	• Global with some overlap in work hours • Multi-cultural (4 countries) • Multi-organizational – 4 units from same company	• Global with some overlap in work hours • Multi-cultural (2 countries) • Multi-organizational – 2 units from same company	• Global with small overlap in work hours • Multi-cultural (2 countries) • Multi-organizational – 2 units from same company
Specifics of the engineering task	• Customer project specific engineering lasting 6-12 months	• Open-ended participation in product development projects • Software development is the primary activity	• Customer project specific engineering lasting 3-6 months	• Customer project specific engineering lasting 3-6 months
Operational set-up	• GEN lead based at customer location • Use of groupware (Lotus Notes) • Supplier does "simple" work	• GEN lead at IC location • Use of groupware (Lotus Notes)	• GEN lead at IC location • Use of groupware (Lotus Notes) • Shared design tool • EC location for mechanical design	• GEN lead at IC location • Use of groupware (Lotus Notes) • Common CAD tool • EC location for electrical drafting
Method used for case data collection	• Author was part of the GEN team – during set-up and operation	• Involved in set-up of the GEN team and in periodic review of the operations	• Author had no direct involvement and information was gathered via interviews with key managers	• Author set up this GEN

manpower build-up. At one of the IC locations a considerable amount of time was invested by engineers to train the engineering staff at the supplier.

The GEN challenges were many. The geographical distance between North America, Europe and Asia made it hard for people to have frequent face to face meetings and rapidly gain trust with each other. Overlapping work brought people in early hours in US and late night in China to attend meetings via telephone or video conferencing. Evolution of organization and structure took time, and had to go through resistance in implementation from both IC and EC locations. Commercial and project execution issues needed several iterations before rules of the joint work were clear. Overcoming both technical and regional cultural differences was a major challenge. Technical differences arose in the case of interpreting and understanding engineering tasks and outcomes. For example, quality and safety were interpreted differently in different countries. Other differences arose from languages, historical background, and regional cultures. For example, style of speech and accent were early obstacles before the precise meaning a person was implying could be understood. The strategy for the GEN transition and implementation was fraught with errors due to the fact that this was being done for the first time in this business. Clarity in setting up the organization, commercial rules of engagement, and formal change management implementation are some of areas where re-work was needed from learning-by-doing. Some key elements of the strategies adopted in transition as per GEN transition model are presented in Table 11.2.

The transition in IC locations was driven as both "top-down" and "bottom-up" approaches. From top-down, management made clear the rationale for engineering in EC and implemented a result-oriented progress tracking. Managers in charge of this process were given clear financial targets for the work they will carryout in EC. From bottom-up, joint project work packages were created to make people work together from simpler to complex engineering tasks.

The competence build-up in EC was managed by series of exchanges of people from both EC and IC countries on joint projects for customers or for internal R&D. The R&D projects were used to standardize and simplify fundamental engineering solutions that can be used across customer projects. Thereby, resources in EC quickly became capable of engineering customer projects using these standards. The process of standardization is described in detail in Nidamarthi and Karandikar (2005).

Team building was the most abstract of the three transitions. At the start, this was not identified and thereby no one explicitly owned or managed this transition.

Table 11.2. *Model strategies adopted in Case 1*

Foundation elements 1. Clear allocation of work between locations made with the EC locations charged with low complexity design to begin with. 2. The portfolio of designs standardized to reduce extent of re-engineering work.
Transformation processes 1. All units given financial targets by global management for the amount of the engineering work they will carry out in ECs. 2. The engineering head of EC unit selected from one of the IC business units and transferred there for a 2-year period.

Eventually it became clear and the global head of this business took responsibility to manage the team building across countries. Clear allocation of work, gradual competence build-up from standards based engineering to complex tasks, and review of change management with the managers in IC were the major parts of this transition.

This particular engineering network is past the transition stage. One consequence of the set-up of this GEN has been that the detailed project engineering activities at one of the three IC locations have been eliminated. Furthermore, the build of competence at the EC location has progressed well though it took much longer than initially planned.

11.6.2 Case 2 – Power Automation Business

The Power Automation business develops and sells products and solutions for protection, control, automation, and monitoring of power systems. Software development is a significant part of product development in this business and it covers software that is embedded in the products as well as software applications for the customer project specific configuration of these products. Due to the sensitive nature of the product applications the software has to have very high quality and reliability.

The principal motivation for setting up a global engineering network for Case 2 was the lack of sufficient numbers of competent technical resources in the IC locations and the pressure for getting the new products to market by the planned deadlines. Thus delaying product launches to accommodate fewer IC resources was not an option. The possibility to reduce development costs was seen as an added bonus. Subsequently, a reduction in the total global budget for the development projects without a change in the scope of the work necessitated the use of additional EC resources. Some key elements of the strategies adopted in transition as per GEN transition model are presented in Table 11.3.

Table 11.3. *Model strategies adopted in Case 2*

<u>Foundation elements</u> 1. Test activities started in EC unit using work packages that could be easily "outsourced" from the IC locations, e.g., reverse engineering of a piece of software. The suitability of the task at hand for distribution in a global development team is an important consideration in the organization of effort. 2. The software development and life cycle management tools used at all locations are common which makes the sharing of effort easy.
<u>Transformation processes</u> 1. Cross-fertilization of the network members initiated. Managers from IC location visited the EC location and developers for EC location were stationed at the IC location for up to 3 months. 2. Once trust in the competence of the EC unit personnel was built up these resources were integrated more tightly into the strategic product development planning of the IC units.

11.6.3 Case 3 – Power Products Business

This is a case of an engineering-to-order electrical equipment business with an annual product volume of a few hundreds. The customer base of this business is worldwide and for this particular range of products the principal engineering center was in a high wage country in Western Europe. The high demand for the products in the ECs where cost is a critical factor and a gradual aging of the engineering department at the IC location coupled with the need to reduce project engineering costs provided the motivation to seek resources in an EC. The decision to set up an engineering resource pool in the EC factory of the same company was made in 2004.

There was significant initial resistance to this strategy in the IC location engineering department. However, it turned out to be less of a factor since the business in the EC market was booming and the need to expand there was obvious as well as the fact that other assignments were found for the "excess" engineering staff at the IC location. Thus no redundancies were necessitated.

Some key elements of the transition strategy are presented in Table 11.4. The GEN is beyond the transition phase today. The planned resource build-up at the EC location and their training has been accomplished. The nature of the work in the network is now truly collaborative. The availability of global standard design rules, implemented in a globally rolled-out engineering software application that could generate custom designs for local markets, was a huge positive factor in the successful start of activities in this GEN.

11.6.4 Case 4 – Power Products Business

As with Case 3, this is a case of an engineering-to-order electrical equipment business with an annual product volume of a few hundreds. The customer base of this business is mainly North America and thus the principal product development as well as engineering center was in the United States. An increase in project volume due to a booming market coupled with the need to reduce project engineering costs provided the motivation to seek resources in EC. A formal evaluation process was set up to evaluate at least three possibilities for EC resources, two using internal company resources in EC locations and one using a third party service provider.

Table 11.4. *Model strategies adopted in Case 3*

<u>Foundation elements</u> 1. There was a clear organization of effort that led to a decision to allocate mechanical design activities to the EC location. 2. Highly standardized global design rules for the product were developed and enforced by embedding them in a common design tool whose use was mandatory at all locations.
<u>Transformation processes</u> 1. Three year plan made for resource build-up at EC location. This was coupled with engineering workforce reduction (through retirements and reassignment of existing staff) over the same time period. 2. A training program was defined that brought over the EC staff to the IC location for 3–6 months to learn the engineering practices while working on actual projects.

Due to the fact that considerable product know-how would be needed to support this activity the third party provider was quickly ruled out. The selection criteria used for the two remaining internal EC locations included the following:

1. Availability of sufficient numbers of competent engineers in the EC location.
2. Knowledge of technology/product.
3. Understanding at the EC location(s) of IC location procedures and processes.
4. Labor cost (hourly as well as annual).
5. Commonality of engineering tools between the locations.
6. Potential for IC location to learn from the EC location.
7. Potential for cycle time reduction.
8. Quality of IT infrastructure for collaborative work.
9. Ease of communication – infrastructure, English language skills and work culture fit.
10. Commitment of local management.
11. Alignment with global business strategy.

These criteria are also reflected in the globalization value proposition described in Dehoff and Sehgal (2007). There was minimal resistance to this strategy in the IC location engineering department. Since the business in the EC market was booming and the need to expand there was obvious. Additionally, it helped that that other assignments were clearly identified for likely "excess" engineering staff at the IC location and no redundancies were foreseen.

Some key elements of the strategies adopted in transition to a GEN are presented in Table 11.5. This GEN is just moving beyond the transition phase. The resource build-up at the EC location and their training has been planned and full scale collaborative activities initiated. The set-up of this GEN once again highlighted the importance of documented standard design rules and a common and shared engineering tool. Another important point is that while this GEN would be categorized in the Eppinger and Chitkara (2006) model as Captive Offshore it was contractually set up as a Global Outsource model. This was deemed necessary in order to instill greater confidence at the IC location in the operation of this GEN. This was a result of the explicitly stated accountability, in the contract agreement, of the teams at each location in the network.

Table 11.5. *Model strategies adopted in Case 4*

Foundation elements 1. There was a clear organization of effort that led to a decision to allocate electrical drafting activities to the EC location. 2. Use of a common design tool was mandated at the EC location. 3. A formal contract was established between the IC & EC location.
Transformation processes 1. Long term plan made for resource build-up at EC location with plans to update on a yearly basis. 2. The key person from the EC locations spent 8 weeks at IC location and key staff from IC location were required to visit the EC location. A two-week follow-up visit was done in Year 2. 3. The EC staff executed a few test projects – both off-site and on-site to test drive the collaboration process.

11.7 Concluding Remarks

In this chapter we briefly discussed issues that have been dealt with in research concerning the operation of global, multi-cultural, multi-organizational, virtual teams. Such teams are becoming the norm for most manufacturing companies when it comes to new product development or project-driven engineering activities. While many companies have such teams in operation many are still in the process of making such a transition. There is a strong trend towards using resources from ECs. The rationale for this is usually the non-availability of sufficient competent resources at one location, the need to shorten development times or the necessity to reduce development and engineering costs. Our focus here has been specifically on GEN that has membership from IC and EC locations as these are increasingly common today and that pose additional challenges as compared to GEN with only IC membership. Further, the challenges of making the transition to GEN have been emphasized as the process is fraught with business risks. The challenges of IC and EC membership and the difficulties encountered during the transition phase have not been explicitly covered in existing literature.

A model for managing the transition was described. The author's involvement in the transition process of the cases presented here led to the formulation of the transition model. Subsequently, observations made along the way, particularly with respect to the problems encountered, led to the refinement of the model. The model has certain foundational elements and is based on the execution of three location-dependent transformation processes, namely, change management, team building, and competence building. The transition process, which can take a few months, when managed properly and explicitly with appointed owners for the three transformations can lead to significant benefits due to tapping of a new pool of technically competent resources and increased first-hand knowledge of customer behavior in the new markets. As of this writing, a successful transition to GEN has been made in all cases presented here. One common aspect critical to this success was the fact that in all cases cost reduction was not made the sole and primary driver to set up the GEN. It was seen as a result or an added benefit. Additionally, the practice of "closing the loop" is being assiduously followed within each network primarily through the frequent use of verbal, i.e., telephonic, conversations.

The ideas presented here were motivated by the fact that there is a lack of literature addressing issues faced in transition to a GEN. Further research is also needed concerning the mechanisms for and the obstacles to the reconfiguration of GENs. Once again, most literature that focuses on the operation of global teams, with some exceptions such as Dekkers and van Luttervelt (2006), assumes a steady state operation. However, these teams go through frequent changes in terms of individuals as well as organizations. The criteria for successful management of such reconfiguration, sharing many commonalities with the GEN transition process discussed here, needs further research. This need is higher in the case of engineering as opposed to manufacturing as the process and need for know-how and knowledge transfer is more challenging.

The elements of the transition model that is presented establish the necessary conditions to facilitate a trouble-free transition to a GEN. However, further research will be necessary to determine if the conditions established by these elements are also sufficient to facilitate a trouble-free transition. If not, the model may need further enhancement.

References

Benson-Armer, R. and Hsieh, T-Y. (1997), "Teamwork across time and space", The McKinsey Quarterly, No. 4, pp. 18–27.

Bevan, J. (1989), "Co-Makership", Management Decision, Vol. 27, No. 3, pp. 50–54.

Bloch, M. and Jans, C. (2005), "Reducing risks in offshoring projects", The McKinsey Quarterly, No. 3, pp. 10–11.

Boutellier, R., Gassmann, O., Macho, H. and Roux, M. (1997), "Management of Dispersed Product Development Teams: The Role of Information Technologies", R&D Management, Vol. 28, No. 1, pp. 18–27.

Couto, V., Mani, M., Lewin, A.Y. and Peeters, C. (2006), "The Globalization of White Collar Work – Facts and Fallout of Next Generation Offshoring", Booz Allen Hamilton Viewpoint, November.

Dehoff, K. and Sehgal, V. (2007), "Innovators Without Borders webinar: Globalization of Engineering Study", Booz Allen Hamilton, July 17.

Dekkers, R. and Luttervelt, C.A. van (2006), "Industrial networks: capturing changeability?", International Journal of Networking and Virtual Organisations, Vol. 3, No. 1, pp. 1-24.

Dietz, M.C., Lin, S.S.-T. and Yang, L. (2005), "Protecting intellectual property in China", The McKinsey Quarterly, No. 3, pp. 6–8.

Earley, P.C. and Gibson, C.B. (2002), Multinational Work Teams: A New Perspective, Lawrence Erlbaum Associates, Mahwah, NJ.

Eppinger, S.D. and Chitkara, A.R. (2006), "The New Practice of Global Product Development", MIT Sloan Management Review, Vol. 47, No. 4, pp. 22–30.

Grabowski, H., Lossack, R.-S., Gebauer, M., and Hornberg, O. (2001), "Distributed Knowledge Management – New Challenges for Global Engineering and Product Creation: Experiecnes from Chinese-European Collaborations", Proceedings of ICeCE 2001, Sept. 16–18, Xi'an, P.R.China.

Hoare, R. and Seiler G. (2001), "Building Blocks for Capital Projects", The McKinsey Quarterly, No. 2, pp. 56–63.

Jonsson, N., Novosel D., Lilliesköld, J. and Erikson, M. (2001), "Successfull Management of Complex, Multinational R&D Projects", Proceedings of the 34th Annual Hawaii International Conference on Systems Sciences (J. Ralph and H. Sprague, Eds.), Maui, Hawaii, IEEE Computer Society Press.

Karandikar, H. and Nidamarthi, S. (2006), "A Model for Managing the Transition to a Global Engineering Network Spanning Industrialized and Emerging Economies", Journal of Manufacturing Technology Management, Special issue on Dispersed Manufacturing Networks, Vol. 17, No. 8, pp. 1042–1057.

Lee W.B. and Lau H.C.W. (1999), "Factory on demand: the shaping of an agile production network", International Journal of Agile Management Systems, Vol. 1, No. 2, pp. 83–87.

Mar-Yohana, M.J. (2001), "Leading Engineering Teams in a Global Environment", 31st ASEE/IEEE Frontiers in Education Conference, October 10–13, Reno, NV.

McDonough, E., Kahn, K., and Barczak, G. (2001), "An Investigation of the Use of Global, Virtual, and Collocated New Product Development Teams", The Journal of Product Innovation Management, Vol. 18, No. 2, pp. 110–120.

Nidamarthi, S. and Karandikar, H. (2005), "Architecting and Implementing Profitable Product Families and Shared Engineering Platforms Subject to Organizational Constraints", in: Product Platform and Product Family Design: Methods and Applications, (eds. T. W. Simpson, Z. Siddique and J. Jiao), Springer, New York, pp. 475-498.

Nowak, T., Banas, M., Florkowski, M., Fulczyk, M. and Karandikar, H. (2002), "Synchronous collaborative design system - functionality and architecture", Proceedings of the 9th ISPE International Conference on Concurrent Engineering: Research and Applications, July 27–31, Cranfield.

Parker, H. (1999), "Analysis of Collaboration in the New Product Development Process", Proceedings of the 10th Annual Conference of the Production and Operations Management Society, POM-99, March 20–23, Charleston, SC.

Petersen, K.J., Handfield, R.B. and Ragatz, G.L. (2003), "A Model for Supplier Integration into New Product Development", Journal of Product Innovation Management, Vol. 20, No. 3, pp. 284–299.

Powell, A., Piccoli, G., and Ives, B. (2004), "Virtual teams: A review of current literature and directions for future research", ACM SIGMIS Database, Vol. 35, No. 1, pp. 6–36.

Ragatz, G.L., Handfield, R.B. and Scannel, T.V. (1997), "Success Factors for Integrating Suppliers into New Product Development", Journal of Product Innovation Management, Vol. 14, pp. 190–202.

Reddy, Y.V.R., Srinivas, K., Jagannathan, V. and Karinthi, R. (1993), "Computer Support for Concurrent Engineering", IEEE Computer, Vol. 27, No. 1, pp. 12–16.

Snow, C.C., Miles, R.E. and Coleman, Jr., H.J. (1992), "Managing 21st century network organizations", Organizational Dynamics, Vol. 20, No. 3, pp. 5–20.

Stock, G.N., Greis N.P. and Kasarda J.D. (1999), "Logistics, strategy and structure: A conceptual framework", International Journal of Physical Distribution & Logistics Management, Vol. 29, No. 4, pp. 22–23.

Vestring, T., Rouse, T., Reinert, U. and Varma, S. (2005), "Making the move to low-cost countries", Supply Chain Strategy, Vol. 1, No. 2.

Vinaja, R. (2003), "Major Challenges in Multi-cultural Virtual Teams", Proceedings of the conference of the American Institute for Decision Sciences (Southwest Region), March 6–8, Houston, Texas, USA, pp. 341–346. Also accessible on-line at http://www.sbaer.uca.edu/research/swdsi/2003/Papers/068.pdf.

Zhang, Y., Gregory, M. and Shi Y.J. (2007), "Global engineering networks: the integrating framework and key patterns", Proc. IMechE, Vol. 221, No. 8, Part B: Journal of Engineering Manufacture, pp. 1269–1284.

EPILOGUE

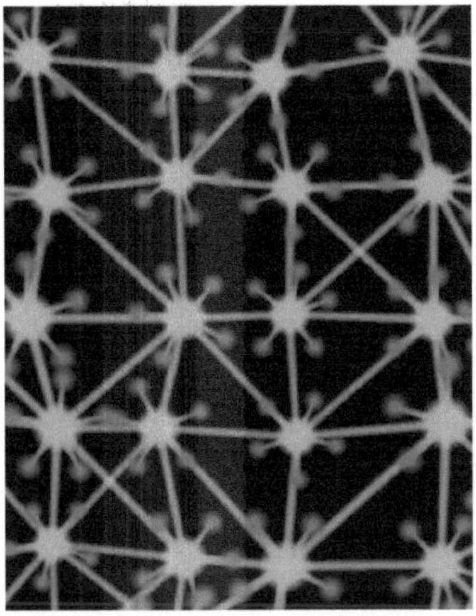

12 What Follows ...

This edited book has exposed a wide range of contributions, from empirical to theoretical, addressing the issues surrounding Dispersed Manufacturing Networks. The contributions neither claim to be complete nor will be the final stage of development of theory, methods and tools for industrial networks that consist of more loosely connected entities; rather they show the continuous evolution from our views on manufacturing networks (for some explicit comments on how these views have changed, see Chapter 2 by Rob Dekkers and David Bennett). It must be noted that the necessity to look at organisations that are loosely connected dates back to works like that of Weick (1976); he states that the perspective of loosely connected entities incorporates a surprising number of disparate observations about organisations, suggests novel functions, creates stubborn problems for methodologists, and generates intriguing questions for scholars. However, this recognition has hardly resulted in approaches that describe these loosely connected entities as networks; most authors (like Dubois and Gadde [2002] for the construction industry) attribute features of loosely coupled entities to specific networks. Mayntz (1993) has, much like the position taken in Dekkers and van Luttervelt (2006, p. 4) and Dekkers *et al.* (2004, pp. 65, 71–73), recognised that networks of loosely connected entities might solve coordination problems of responsiveness, agility and variety. This book will conclude by briefly looking back at the contributions made related to these unique capabilities of Dispersed Manufacturing Networks and the challenges that further research needs to address for loosely coupled manufacturing networks (Section 12.1) and the implications the thinking and research has on practice (Section 12.2).

12.1 Contribution by This Book

What follows from the contributions is that contemporary manufacturing networks with loosely connected entities have come about through two mechanisms. First of all, the manufacturing networks have emerged as a result of collaboration between loosely connected entities (so-called Collaborative Networks [Camarinha-Matos and Afsarmanesh, 2005, p. 439]). That mostly concerns SMEs that coordinate either globally (Subsection 2.3.1) or regionally (Chapter 3 by Hamid Noori and W.B. Lee) but with the explicit aim to have a wider reach; the latter resembles the regional networks labelled "Third Italy" by Biggiero (1999) and Robertson and Langlois (1995, p. 549). The second source of manufacturing are the global production networks that come about through OEMs (Chapters 8 by Joachim Kuhn and 9 by Stephen Smith *et al.*), as similarly described by Ernst (2002) and Sturgeon (2002). The characteristics of global production networks correspond to those of Strategic Networks, as discussed in Section 1.1. The challenges these two forms of more loosely connected organisations as networks face, bring us back to the questions raised at the end of Section 1.2:

- *Which forms of industrial networks emerge and meet the industrial challenges of this century? Which performance criteria do they meet?*

  The book has identified two origins of industrial networks that have emerged: Collaborative Networks and global production networks. The concept of loosely connected entities characterises these networks as agents that share common objectives, e.g. addressing market needs, and pool resources to have a wider reach, e.g. global competitiveness. At the same time, these networks do not operate in a rigid structure but allow flexible allocation of resources, e.g. planning and scheduling to reflect actual demand, and reconfiguration (e.g. inclusion and exclusion of members based on continuous evaluation); Miles and Snow (1984, p. 19) call these dynamic networks. These dynamic networks should increase customer responsiveness and also enhance the innovative capabilities exceeding the capabilities of individual firms. The dynamic behaviour along these two dimensions classifies the networks as Complex Adaptive Systems (see Chapters 3 by Hamid Noori and W.B. Lee and 5 by Rob Dekkers; a further expansion of the complexity perspective is forthcoming in Dekkers [2009]). This aligns the Dispersed Manufacturing Networks with the concept of dynamic capabilities (Teece *et al.*, 1997), as mentioned by Colotla *et al.*, 2003, p. 1203), albeit that the development of capabilities takes place at the network level and the level of the individual entity. The developments at both levels then calls for the identification and allocation of private and common benefits, see Subsection 5.3.5; a topic that is under-researched given its impact on the instability of industrial networks. Hence, these loosely connected entities as networks, so-called dynamic networks, display the non-linear behaviour of Complex Adaptive Systems.

  This non-linear behaviour enhances responsiveness and innovative capabilities of the individual firms and the network. Within supply networks the responsiveness is generally denoted with agility (e.g. Goldman and Nagel, 1993, p. 31). Particularly Chapters 6 (by Petri Helo *et al.*) and 7 (by Hossein Sharifi *et al.*) have added to concepts for agile supply networks by providing open source software for control between companies and a framework for design of agile supply chains in relation to the dynamics of growth and diversification. In addition, collaboration with other companies has a significant impact on the dynamic capabilities of a network, which resembles the concept of changeability (Milberg and Dürrschmidt, 2002); they denote the changeability as the sum of flexibility, the capability to operate in a wider window on dimensions of business management, and responsiveness, the ability to handle emerging changes of the environment. Thus, the changeability indicates the total changes a network absorbs (Wiendahl and Lutz, 2002); Jahre and Fabbe-Costes (2005, p. 153) follow a similar reasoning. Changeability then calls for reconfiguration, either by self-similarity based on fractals (Section 4.4) or by optimisation of the organelle structure (Subsection 2.3.1); the basis for these reconfiguration approaches is similar: the integration of business processes: material flow and information flows. That implies as Dubois and Gadde (2002, pp. 628–629) state that these loose couplings stimulate the innovative potential of the entities in the network and the networks itself (that aligns with the observed innovative potential and the changeability of and adaptation by the Italian districts [Biggiero, 1999]).

- *Which models that will lead to methods and tools do apply for networks? How about the coordination and inter-organisational integration in networks of loosely connected agents?*

  For those loosely connected networks, the key to managing the business processes is the monitoring of the capability of individual participating entities (Section 3.2, 3.7 and Section 4.4); this is called self-criticality by Hermann Kühnle in Section 4.2 and it also appears as concept in Section 3.5. It comes back in the central role of hubs that enabled by information and communication technologies (Section 3.5) exert that capability; the distributed plant automation, PADABIS, is an example (Section 4.5), based on the notion of Spaces of Activity. That capability facilitates learning of the network and adaptation to changing circumstances; it strongly resembles the concept of process capability in the steady-state model that is mentioned by Dekkers (2005, p. 431). This then calls for reconfiguration, either by self-similarity based on fractals (Section 4.4) or by optimisation of the organelle structure (Subsection 2.3.1); note that the basis for those reconfiguration approaches – the integration of business processes: physical flows and information flows – is the same. Therefore, the self-criticality in relation to reconfiguration constitutes a core capability of industrial networks and might be even the dynamic capability.

  In this perspective of the dynamic capability, the self-criticality in relation to reconfiguration does not abolish the need for coordination. To that end, Section 7.4 links the supply chain design to the growth strategies of Ansoff, demonstrating that coordination in the supply chain depends on the variety and flexibility of products offered. That design approach evaluates supply chains from a holistic perspective, as also advocated by Fisher (1997) and Jahre and Fabbe-Costes (2005). That the coordination has changed by the international dimension due to geographical dispersion manifests itself in Chapters 8 and 9. The contribution by Joachim Kuhn in Chapter 8 clearly demonstrates the impact of decision-making: creating global production networks (see also Ernst [2002] on global production networks and Kim and Lee [2001] on the experiences of Korean carmakers). Joachim's study stresses that the development of global manufacturing capabilities (as identified by Shi and Gregory [1998, p. 209]) has been going on in the automotive industry since the beginning of the 20th century. That is articulated in Chapter 9, where Stephen Smith *et al.* demonstrate that dispersed manufacturing capacity is a common phenomenon, not restricted to the devolution by larger Chandlerian firms. The supply chain networks that emerge from these developments experience issues for control and coordination, as Mahendrawathi Er and Bart MacCarthy show in Chapter 10. Specifically the product variety, often seen as a stronghold of Dispersed Manufacturing Networks and agile supply

chains, amplifies demand and supply uncertainties. Chapter 10 is certainly an addition and expansion to the works of Fisher (1997) and Lee (2002); in fact, it substantiates the propositions made in those papers. This implies that although loosely connected networks have become common ground, they also come along with considerable challenges for coordination.
Additionally, in Chapter 11, the evidence by Harsh Karandikar supports that Global Engineering Networks bring significant benefits due to tapping of a new pool of technically competent resources and increased first-hand knowledge of customer behaviour in the new markets; those benefits exceed possible cost reductions. In that chapter it emerges that the transformation to these engineering networks requires effective transition processes driven by (i) breaking down any resistance to the spread and transfer of development activities to locations in emerging countries, (ii) building up a strong team and focusing on improving the level of trust and increasing the collaboration among the various sites, and (iii) focusing on building up competence and expertise at the new development sites in emerging countries. Chapter 11 confirms that the concept of networks constitutes a transformation process in approaches to managing loosely connected entities.

- *How should companies manage industrial networks given their instability and the dynamics of the environment? Which routes does management science have to explore?*

All evidence from the chapters in this book only underlines that approaches to management, collaboration and coordination in these networks of loosely connected entities cannot be taken from the Strategic Network perspective (Section 1.1); that would result in issues of power and trust (Huemer, 2004), completely detrimental to the potential benefits of industrial networks. Rather, this book highlights that coordination and control in these networks should rely on either centralised control by hubs (facilitated by transparent decision-making in the hub [Chapter 3] and adequate supply chain management [Chapters 6 and 10]) or negotiation between entities (Chapter 4) and continuous performance evaluation (Chapters 3 and 4) or by new approaches like the distribution of private and common benefits (Chapter 5). At the same time, exploiting the benefits of industrial networks calls on transformation processes (Chapter 11) and appropriate design of coordination in networks (Chapters 6 and 7).
All chapters indicate that the work is not done, yet. In fact, almost all authors point to further research, e.g. models for network reconfiguration (Chapter 2), dynamic descriptions for behaviour in networks (Chapters 2 and 5), information and communication technology exceeding ERP to support networks (Chapters 3 and 4), extension of agile supply networks to a variety of industries (Chapters 6 and 7) and effects of product variety on supply chains (Chapter 10). In addition to the methodologies, methods and tools found in this book, the work will continue to expand the thoughts, methods and tools.

To benchmark the work done so far in this book, it is necessary to evaluate the concepts presented so far; a few words will only be devoted to this evaluation. In his work, Bacharach (1989, p. 510) mentions criteria for evaluating concepts for organisations, which will be applied to the contributions in this book. One dimension is the falsifiability and the utility of methods and tools. On that dimension the contributions in this book introduce a wide variety of concepts that apply potentially to many different contingencies (i.e. utility). The falsifiability is found in the fact that traditional approaches residing in the monolithic company have limited reach for industrial networks consisting of loosely connected entities. First of all, more research should be done before the falsifiability of concepts for networks can be investigated. The second dimension describes methods in terms of variables, constructs and relationships. All across the chapters, the contributions point to models and methods that demonstrate underpinning variables, constructs and relationships. The specific issues and constructs for industrial networks also underpin the call of Camarinha-Matos and Afsarmanesh (2005) that this constitutes a new scientific discipline. Hence, it might be concluded that this book has made a constructive contribution to concepts for organisations, that concepts and constructs have validity for the specific domain of industrial networks and, finally, that the research present in the chapters has resulted or will later yield methods for industrial networks with loosely connected entities.

12.2 Implications for Practice

What follows is that for managerial practice the emphasis on industrial networks requires a change in mind-set from three perspectives. Firstly, the concepts embedded in the thinking about networks as an extension of the monolithic company will yield only marginal benefits. Besides it carries the danger that this management approach will result in issues of power and trust for industrial networks (see e.g. Huemer, 2004), much like the thoughts of the Strategic Network perspective and Resource-Based View (Section 1.1). Otherwise, the management of networks might suffer from fragmentation and its impact on decreasing the effectiveness of networks, as is so characteristic for the construction industry (Dubois and Gadde, 2002); even though others take a contrasting position (Dorée and Holmen, 2004). Secondly, the distribution of private and common benefits needs attention, where traditionally pricing and costs are focus of managerial attention. Although for part, it resembles the embeddedness in networks (e.g. Uzzi,1997, pp. 54, 61), it does not imply that companies need to sacrifice. Rather they might benefit from the increased reach and responsiveness that the networks offer on the long-term, albeit again through different mechanisms than traditional methods applicable for the Chandlerian or monolithic firm. Thirdly, in some sense, collaboration in networks has put smaller and bigger companies at equal footing. That implies that both smaller and bigger companies compete at a global scale with a greater flexibility and changeability. That in itself has accelerated the necessity to operate within networks: the emergence of networks going hand in hand with the necessity. Henceforth, networks have become a reality for many companies. Despite the changes these three perspectives bring about, methods and tools have not fully been settled, this way calling also on managers to contribute to further insight and to collaborate with academics to advance both practice and theory.

References

Bacharach, S.B. (1989) Organizational Theories: Some Criteria for Evaluation, Academy of Management Review, Vol. 14, No. 4, pp. 496–151.

Biggiero, L. (1999) Market, hierarchies, networks, districts: A cybernetic approach, Human Systems Management, Vol. 18, No. 2 pp. 71–86.

Camarinha-Matos, L.M. and Afsarmanesh, H. (2005) Collaborative Networks: a new scientific discipline, Journal of Intelligent Manufacturing, Vol. 16, No. 4–5, pp. 439–452.

Colotla, I., Yongjiang, S. and Gregory, M.J. (2003) Operation and performance of international manufacturing networks, International Journal of Operations & Production Management, Vol. 23, No. 10 pp. 1184–1206.

Dekkers, R. (2009) Distributed Manufacturing as Co-Evolutionary System, International Journal of Production Research, Vol. 47, No. 8, pp. 2031-2054.

Dekkers, R. and Luttervelt, C.A. van (2006) Industrial networks: capturing changeability?, International Journal of Networking and Virtual Organisations, Vol. 3, No. 1 pp. 1–24.

Dekkers, R., Sauer, A., Schönung, M. and Schuh, G. (2004). Collaborations as Complex Systems, in Proceedings of the Designing and Operating Global Manufacturing & Supply Networks, 9th Annual Cambridge International Manufacturing Symposium, IMNet/CIM, Cambridge, 25–26 September.

Dorée, A.G. and Holmen, E. (2004) Achieving the unlikely: innovating in the loosely coupled construction system, Constrution Management and Economics, Vol. 22, No. 8, pp. 827–838.

Dubois, A. and Gadde, L.-E. (2002) The construction industry as a loosely coupled system: implications for productivity and innovation, Constrution Management and Economics, Vol. 20, No. 7, pp. 621–631.

Ernst, D. (2002) Global Production Networks and the Changing Geography of Innovation Systems. Implications for Developing Countries, Economics of Innovation and New Technology, Vol. 11, No. 6 pp. 497-523.

Fisher, M.L. (1997) What Is the Right Supply Chain for Your Product?, Harvard Business Review, Vol. 75, No. 2, pp. 105–117.

Goldman, S.L. and Nagel, R.N. (1993) Management, technology and agility: the emergence of a new era in manufacturing, International Journal of Technology Management, Vol. 8, No. 1/2, pp. 18-38.

Huemer, L. (2004) Balancing between stability and variety: Identity and trust trade-offs in networks, Industrial Marketing Management, Vol. 33, No. 3, pp. 251–59.

Jahre, M. and Fabbe-Costes, N. (2005) Adaptation and adaptability in logistics networks, International Journal of Logistics: Research and Applications, Vol. 8, No. 2, pp. 143–157.

Kim, B. and Lee, Y. (2001) Global Capacity Expansion Strategies: Lessons Learned from Two Korean Carmakers, Long Range Planning, Vol. 34, No. 3, pp. 309–333.

Lee, H.L. (2002) Aligning Supply Chain Strategies with Product Uncertainties, California Management Review, Vol. 44, No. 3, pp. 105–120.

Mayntz, R. (1993) Modernization and the Logic of Interorganizational Networks, Knowledge and Policy: The International Journal of Knowledge Transfer and Utilization, Vol. 6, No. 1, pp. 3–16.

Milberg, J. and Dürrschmidt, S. (2002) Planning Methodology for Changeable Logistic Systems, Production Research, Vol. 9, No. 1, pp. 63–68.

Miles, R.E. and Snow, C.C. (1984) Fit, Failure and the Hall of Fame, California Management Review, Vol. 26, No. 3 pp. 11–28.

Robertson, P.L. and Langlois, R.N. (1995) Innovation, Networks, and Virtual Integration, Research Policy, Vol. 24, No. 4 pp. 543–562.

Shi, Y. and Gregory, M. (1998) International manufacturing networks-to develop global competitive capabilities, Journal of Operations Management, Vol. 16, No. 2, pp. 195–214.

Teece, D.J., Pisano, G. and Shuen, A. (1997) Dynamic Capabilities and Strategic Management, Strategic Management Journal, Vol. 18, No. 7, pp. 509–533.

Uzzi, B. (1997) Social Structure and Competition in Interfirm Networks: The Paradox of Embeddedness, Administrative Science Quarterly, Vol. 42, No. 1, pp. 35–67.

Weick, K.E. (1976) Educational Organizations as Loosely Coupled Systems, Administrative Science Quarterly, Vol. 21, No. 1, pp. 1–19.

Wiendahl, H.-P. and Lutz, S. (2002) Production in Networks, Annals of the CIRP, Vol. 51, No. 2, pp. 1–14.

About the Editor

Rob Dekkers is Reader at the University of the West of Scotland (which he joined in 2006), after having been a Senior Lecturer in Industrial Organisation and Management at Delft University of Technology since 1992. Before that, he worked in industry as internal consultant, production manager and senior project manager. He holds a Master's degree in Mechanical Engineering and a doctoral degree, both from Delft University of Technology. He has authored and co-authored about 80 publications. He serves on several review panels and committees, e.g. the International Foundation for Production Research and the (International) Review Panels for EPSRC's Innovative Manufacturing Programme. The main areas of research cover innovation and technology management, changes and transitions in companies, manufacturing strategy, outsourcing models and industrial networks, underpinned by systems theories, the science of complexity and evolutionary (biological) models.

About the Authors

David Bennett is Professor of Technology Management at Aston Business School, Birmingham, UK. He has various research interests within the fields of Management of Technology and Operations Management. External appointments include an Adjunct Professorship with the University of South Australia. His industrial experience includes periods in the automotive components and electrical equipment industries. David Bennett holds MSc and PhD degrees from the University of Birmingham. He is a Chartered Engineer in the UK and is currently serving as a member of the Board of Directors of the International Association for Management of Technology.

Mahendrawathi Er received a first degree in Industrial Engineering from Institut Teknologi Sepuluh Nopember (ITS), Indonesia. She then obtained an MSc in Operations Management and Manufacturing Systems and a PhD in Manufacturing Engineering and Operations Management, both from Nottingham University. Following her study, she worked as a Research Associate for VIVACE, a large European Union-funded Framework 6 Integrated Project, at Nottingham University Business School. She is currently a Lecturer in the Department of Information Systems at ITS. Her teaching and research interests include supply chain management, international operations, product variety management, modelling and simulation. She has published papers in international journals and presented at conferences.

Petri Helo is a Research Professor and the Head of the Logistics Systems Research Group, Department of Production, University of Vaasa. He is also Visiting Professor with the International Programme of Industrial Engineering at Kasetsart University, Thailand. His research addresses the management of logistics processes and information systems in supply chains, which take place in electronics, machine building, and food industries. Dr. Helo is also partner at Wapice Ltd, a company building industrial software. His publications have appeared in *Industrial Management and Data Systems, International Journal of Production Research, System Dynamics Review,* etc.

Hossam Ismail was educated at the University of Ain Shams, Cairo, where he received the degree of BSc (Hons) in Mechanical Engineering. He obtained his PhD from the University of Birmingham. He is currently a Senior Lecturer at The University of Liverpool Management School. He has publications in the areas of manufacturing systems management, agility and mass customisation. Hossam is Director of the Agility Centre, a research and business support centre to develop agility tools and methodologies to assist manufacturing-based SME's. He has supervised 17 knowledge transfer programmes and he had four major Research and Development projects in design and manufacturing.

Roger J. Jiao is Associate Professor of Computer-Aided Engineering and Design at the Georgia Institute of Technology. Prior, he was Associate Professor at Nanyang Technological University, Singapore, and Lecturer in the Department of Management at Tianjin University, China. He received his PhD from Hong Kong University of Science & Technology, a BSc from Tianjin University of Science & Technology, and a MSc from Tianjin University. His research interests include design theory and methodology, reconfigurable manufacturing systems, operations management, and engineering logistics. His publications appeared in *IIE Transactions, Computer-Aided Design, IEEE Transactions on Engineering Management, Decision Support Systems, International Journal of Production Research*, etc.

Harsh Karandikar is currently responsible for the development of electrical power distribution products for ABB's North American market. Harsh has been with ABB since 1998. Prior to joining ABB, he worked at the Science Applications International Corporation and the Concurrent Engineering Research Center of West Virginia University. Harsh has written extensively on the topic of effective and efficient management of product development including the practice of Concurrent Engineering and the role of product platforms, global teams and emerging country sourcing. Harsh holds MSc and PhD degrees from the University of Houston and a Bachelor's degree from the Indian Institute of Technology, Kanpur, all in Mechanical Engineering.

Natalia Kitaygorodskaya received a PhD degree in Economics and Business Administration from the University of Vaasa, Finland. As a project researcher at the same university, she has been involved in projects concerned with logistics in Finland and Russia. As a member of the logistics research team, she has co-authored several research papers published in peer-refereed journals. Natalia's major field of interest is human interactions and their effects on organisational performance. Her doctoral dissertation was concerned with transactive memory systems in distributed R&D teams. Natalia is also a member of the International Society for Professional Innovation Management.

Joachim Kuhn works at Daimler AG and is responsible for the quality support of foreign production locations manufacturing Mercedes-Benz passenger cars. One of the main tasks encloses non-conformance management, the introduction of sustainable measures including 6σ or Shainin methodologies and operational quality planning. Since 1995, Joachim has been employed with Mercedes-Benz, Sindelfingen plant, starting his career in logistics and in 1998 moving on to the quality area. Also in 1998 he became a Visiting Professor and currently he is Professor of Management Practice in International Logistics and Quality at the Ashcroft International Business School of Anglia Ruskin University (Cambridge campus) while still being employed with Mercedes-Benz.

Hermann Kühnle joined the Otto-von-Guericke-University of Magdeburg, Germany, in 1994 as a Full University Professor for "Factory Operations and Production Systems" and as Executive Director of the Institute for Ergonomics, Manufacturing Systems and Automation. In the years from 1994 to 2001 he was also Foundation and Executive Director of the Fraunhofer Institute for Factory Operation and Automation in Magdeburg. Since 1995, Hermann has been the spokesman for the research field "Advanced Production Systems in Saxony-Anhalt". He is a member of boards of international journals, of companies and of venture capital groups.

Ashraf Labib is a Professor at Portsmouth Business School, and the Associate Dean (Research) of the Faculty. His main research interest lies in the field of OR and Decision Analysis. He is a Fellow of the Operational Research Society, a Fellow of the IET and a Chartered Engineer. He has received the 2008, 2000, and 1999 "Highly Commended Awards" for three published papers from the Emerald Literati Network. He is currently the Associate Editor of *IEEE Transactions on Systems, Man, and Cybernetics*. He has been active in attracting research-funded projects from the EPSRC, the European Commission and industry.

W.B. Lee is the Cheng Yick-chi Chair Professor of Manufacturing Engineering and the Director of the Advanced Technology Manufacturing Research Centre as well as the Knowledge Management Research Centre of The Hong Kong Polytechnic University. He is currently a council member of the Hong Kong Productivity Council and also the ex-President of the Hong Kong Advancement of the Association of Science and Technology. He is a member of the Editorial Board of the *Journal of Engineering Manufacture*. W.B. Lee's research interests include advanced manufacturing technology and knowledge management. He has published more than 300 papers published in referred journals and proceedings of international conferences.

Bart MacCarthy is Professor of Operations Management at Nottingham University Business School. His research spans the analysis, modelling and design of operational systems in business and industry. He has researched and consulted with a wide range of industries. He has had led major research projects over a number of years in supply chain management, planning and scheduling, and mass customisation. He has attracted funding from industry, the research councils and the European Union. He has published widely in Operations Management and Management Science. He is a Fellow of the Institute of Mathematics, the Institute of Operations Management and the Institute of Engineering and Technology.

Hamid Noori is the Laurier Research Chair in Enterprise Integration and Technology Management and Professor of Operations at Laurier School of Business and Economics, Wilfrid Laurier University in Waterloo, Canada. His research interests include technology transfer and commercialisation, network manufacturing and integrated supply chain, and fast cycle product/service design. His publications have appeared in many journals including *Management Science, Naval Research Logistics, IEEE Transactions on Engineering Management, Decision Sciences, International Journal of Production Research, Technovation, OMEGA* and *Interfaces*. He currently holds a Visiting Chair (under the Distinguished Scholars Scheme) at the Department of Industrial and Systems Engineering, Hong Kong Polytechnic University.

David Petty is a Senior Lecturer in Operations Management at Manchester Metropolitan University Business School. Initially trained as a mechanical engineer, on entering the manufacturing industry in 1983, he quickly moved into the field of production planning and control. Prior to entering academia in 2000, he undertook manufacturing systems consultancy in the automotive component sector for 11 years. In particular, he has acted as lead consultant on several successful Enterprise Resource Planning (ERP) implementations. His current research interests are planning and control, supply chain management and the industrial application of information systems.

Iain Reid is Lecturer at the University of Liverpool. Research interests include New Product Development (NPD) with customer-driven manufacturers, mass customisation, Business Process Re-Design, Knowledge Management and e-Quality Management Systems. He originally joined the University in 2002, as the Agility Centre's Project Manager. Iain has successfully worked with over 80 SMEs via a number of ERDF funded programmes. He has also supervised a number of Knowledge Transfer Partnerships, including new process introduction, product development and process optimisation and the introduction of an ERP system. Prior to embarking on an academic career Iain had over 10 years working with OEMs in NPD, production and quality related areas.

Sari Salminen has a MSc (Economics) from the University of Vaasa, Finland. She has worked with the global networks modeling issues at the ABB Corporate Research Center, Vaasa, Finland, and developed a performance measurement framework for supply chains for ABB companies. Currently, She is working with the finance operations of retailing at GNT Finland Ltd.

Hossein Sharifi holds a PhD in Industrial Engineering/ Management from Liverpool University. After several years of working experience in the high tech industry, he followed a career in academia where he started leading research into the subject of agile manufacturing systems. He has since extended his research into a number of areas including agile supply chains, e-Government theories and strategies, and innovation and knowledge transfer management. His role as a research champion in the multi-university Innovation and Productivity Grand Challenge has opened a new horizon of interdisciplinary research into the role and contribution of university innovation and knowledge.

Stephen Smith undertook research at the University of Manchester under the Engineering Doctorate programme. His doctoral thesis focused on the modelling and simulation of extended supply chains and this research was funded by Strix Ltd. He is currently a Senior .NET Developer at Zen Internet.

David Trustrum has worked for Strix Ltd. on the Isle of Man for 16 years since graduating from Loughborough University with a degree in Manufacturing Engineering. Having held various operational and engineering positions within Strix, he moved into the logistics function where he currently holds the position of Logistics Manager.

MIX
Papier aus verantwortungsvollen Quellen
Paper from responsible sources
FSC® C105338

If you have any concerns about our products,
you can contact us on
ProductSafety@springernature.com

In case Publisher is established outside the EU,
the EU authorized representative is:
Springer Nature Customer Service Center GmbH
Europaplatz 3, 69115 Heidelberg, Germany

Printed by Libri Plureos GmbH
in Hamburg, Germany